VOICE, VIDEO, AND DATA NETWORK CONVERGENCE

ARCHITECTURE AND DESIGN, FROM VOIP TO WIRELESS

JUANITA ELLIS

CHARLES PURSELL

JOY RAHMAN

ACADEMIC PRESS

An imprint of Elsevier Science

Amsterdam Boston Heidelberg London New York Oxford Paris
San Diego San Francisco Singapore Sydney Tokyo

Academic Press
An imprint of Elsevier Science
525 B Street, Suite 1900, San Diego, California 92101-4495, USA
http://www.academicpress.com

Academic Press
84 Theobald's Road, London WC1X 8RR, UK
http://www.academicpress.com

Library of Congress Cataloging-in-Publication Data

Ellis, Juanita.
  Convergence of voice, video, and data networks / Juanita Ellis, Charles Pursell, Joy Rahman.
    p. cm.
  ISBN 0-12-236542-9
  1. Convergence (Telecommunication) I. Pursell, Charles. II. Rahman, Joy. III. Title.

TK5101.E48 2003
384.3′3—dc21   2003050271

International Standard Book Number: 0-12-236542-9

PRINTED IN THE UNITED STATES OF AMERICA
03 04 05 06 07   9 8 7 6 5 4 3 2 1

Voice, Video, and Data Network Convergence

WITHDRAWN

To my dad, Charles Ellis.
—Juanita Ellis

To my mother, Zohra Rahman,
and my late father, Shahidur Rahman.
—Joy Rahman

To my mother, and the rest of my family, who
gave me support, encouragement, perspective, and
the occasional constructive needling; and to
the memory of Herbert Pursell, who started
me out on this journey.
—Charles Pursell

CONTENTS

PREFACE

The concept of voice and data networks is everywhere. We are seeing companies implement these technologies in order to cut management costs, provide integrated voice and data applications, and reduce long term network ownership. The legacy model delivers voice, video, and data separately over dedicated, single purpose networks. Unfortunately, it no longer meets the needs of corporations moving forward.

As a result, businesses are looking for more dynamic and flexible services for communications within their organizations and more importantly, communicating with the outside world and their customers. On the other hand, in these economic hard times, companies have been putting emphasis on reducing or maintaining costs. Convergence is a key to providing this balance of reducing costs while providing increased ability to service customers and improve communications internally within the organization.

This book covers convergence of voice, video, and data networks from a vendor neutral perspective. With so many vendors competing in this market, it is important to understand the lingo and true capabilities behind each of these vendor solutions before making a buying decision. Specifically, this book covers detailed information on convergence network models, protocol stacks, routing algorithms, gateways, and switches required to support these networks. Techniques for deploying, testing, and ensuring QoS are also covered in great depth. Some of the key areas include:

- technologies and planning concepts to help the reader understand the lingo and true capabilities behind Convergence.
- latest standards and those that are being developed in this technology field.
- the IP-Based Converged Network Elements and risks and benefits associated with each environment.
- latest standards and implementations of video and data over wireless networks and how convergence will be a driving force.
- provide technologist with the information needed to start selecting the right approach to deploying converge and ensuring an open architecture.
- designing and testing the network to ensure it meets QoS requirements.

Associated Benefits and Risks Factors of Convergence

As with any technology, there are both benefits and associated risks. Being the pioneer can make a company the leader in the market if they are aware of potential risks and can carefully manage these risks. Even with all the benefits of convergence, companies need to move up this path. A clear understanding of the business drivers is crucial prior to investing in technology solutions that may not meet open standards, select vendors that may dissipate, or follow into the hole of technology driving business.

Benefits of Convergence

Convergence provides network and operational cost reductions while allowing companies to utilize more robust integrated applications. Networking costs can be reduced by converging voice and data onto a single local area network (LAN) infrastructure. This has a variety of potential benefits. For example, benefits include eliminating the need for voice line cards and remote line equipment. This can also eliminate separate, duplicated voice cabling to the desktops as occurs with most IP phones sharing the PC's ethernet connection. Operational savings can be realized during adds, moves and changes. In just movers alone, savings between $10 and $40 are realized per move for the typical large enterprise. In the world of converged applications, companies are able to provide innovative CRM (Customer Relationship Management) and unified messaging applications. These blend voice and data to create new value for the enterprise and their customers.

Lower Recurring Transmission Charges

Companies can significantly reduce their monthly phone bills by directing voice calls over the corporate data network rather than through a carrier. These savings, of course, depend on several factors such as the volume of intra-company calls and the distances between company offices. This is especially true for companies with overseas offices in a foreign country that has a highly monopolistic telecom market. Their savings would obviously increase since they could eliminate many of the costly international long-distance charges.

These savings can also include calls outside of the company. This is accomplished by simply doing two things. First, route calls bound for destinations outside of the company over the corporate network to the closest remote office. Second, interface with the public switched telephone network (PSTN) at that point.

The communications technology today challenges business decision-makers with a wide range of new, sophisticated multifunctional solutions to consider. With the rapid development of enhanced services, they are faced not only with compelling opportunities but potentially costly downfalls as well. Convergence intends voice, fax, data and multimedia traffic that are transmitted over a single multipurpose network. The advances in converged networks are appealing from both business and technological standpoints. There are many advantages to combining a company's various types of communication over a common infrastructure. These advantages include lower recurring transmission charges, reduced long-term network ownership costs, and the ability to deploy a wide range of powerful voice-enabled applications.

The Ability to Deploy Powerful New Integrated Voice-And-Data Applications Convergence not only allows businesses to save underlying costs but also extends market-share through more robust applications focusing on call center solutions, unified messaging and real-time collaboration. By using VoIP, for example, site visitors can converse with a call center agent and quickly address any question or problem. Other examples include real-time multimedia conferencing, distance learning, telecommuting and the embedding of voice links into electronic documents. Such real time applications are revolutionizing business transactions as well as interactions with customers and employees.

Economic Factors

The competitive advantages of transmitting voice calls over the data network provide businesses with next generation solutions that are economically appealing. First, the efficiency of bandwidth is improved by compressing the characteristics of human speech such as pauses and therefore, consuming relatively little bandwidth by "piggybacking" on existing data network connections. Secondly, costs of the digitized transport of a call are typically a fraction of the recurring costs charged by carriers to carry that same calling volume.

Reduced Long-term Network Ownership Costs

Traditionally, companies owned two separate networks—one for voice and one for data. Converged network architecture not only reduces monthly phone bills but also reduces the cost of owning and managing two separate networks. One concern, however, is due to the lack of expertise in this new emerging field. There are so few specialists that the demand for technicians skilled in convergence technology has increased salaries considerably, making it difficult to recruit and retain such engineering talent.

Risks Factors of Convergence

Any potential business gain comes with a variety of potential business risks. Converged networks with voice over IP are no exception. Businesses need to carefully evaluate the converged network solution that best integrates with its existing network. Knowing what the drivers are for convergence is perhaps the most important prerequisite to a sound convergence strategy. There are decisions to make between integrating present equipment to keep costs low or replacing it with more open converged platform architecture.

These decisions determine short term costs such as the transition to new equipment, processes and procedures will require additional training and implementation expenses versus long term operational savings.

A deployment strategy is required to assure an immediate competitive advantage. Businesses need to move from clarifying the goals to eliminating the potential obstacles. This is done by paying careful attention to the selection of vendors, products and deployment schedules, and obtaining the desired transparency for the end user. Understanding the specific needs of the users and evaluating the proposed system capabilities against them is critical to avoid significant support and re-training costs. Other considerations to be made regard the support of existing terminals, fax machines, and conferencing services. Naturally, a converged network also requires competent staff capable of monitoring the converged communications system efficiently as well as security services capable of protecting the integrity of voice, video, and data traffic.

Loss of Voice Quality

Data networks are very different from voice networks. On the data network, packets bounce around somewhat randomly. They can collide and get distorted or even lost. Even the millisecond delays don't affect most program applications as there are mechanisms in ethernet hardware and the IP protocol itself that readily compensate for these phenomena on the data side. On the contrary, voice calls require high, real-time quality. Anything less in voice quality can affect the normal way of doing business.

Loss of Reliability

We all know from experience that data networks are not yet as reliable as voice networks. But that's not keeping VoIP technology from exploding into the global market. Companies that have positioned themselves with a reasonable amount of in-house expertise in converged networking will have a cutting edge on servicing customers to take advantage of its explosive growth. Companies, however, that enter into VoIP at that point without any convergence experience will be at a competitive disadvantage. They won't be able to effectively partner with other companies that have made convergence a core component of their technology portfolios. It is therefore essential that any company with intentions to survive and thrive in a communications-based economy take steps now to ensure its ability to compete in a future that will clearly include universal adoption of VoIP technology.

Lack of Expertise and Experience in Convergence Technologies

How the implementation of VoIP technology impacts the technology professionals needs to be considered carefully. As things stand now, data applications are typically the responsibility of an IT team while voice communications fall under a separate and distinct telecommunications group. The concept of convergence can potentially disrupt the roles and structures of corporate technology staff by posing not only threats to their positions, but forces company executives to consider training as well as possibly retaining skilled experts with convergence experience.

Being Prematurely "Locked-in" to a Given Vendor's Architecture

The rapid pace of change in computing and communications technology today makes vendor "lock-in" a major concern for any potential buyer. Decision-makers are faced with the concern of committing to a particular vendor's product that could potentially narrow their choices in the future. This concern is compounded by the lack of clear standards in the VoIP market. A long-term commitment to one vendor's approach to voice/data convergence may cause a lock-in that extends far beyond the VoIP solution itself and this is what technology managers want to avoid. Without this commitment to standards-based technology, PBX vendors' VoIP solutions won't be a very good bet for companies seeking broad interoperability and flexible migration paths.

Convergence is not a new concept. You may have remembered the earlier IP Telephony applications that were available on the market and, in some cases, free via the Internet. These voice based applications had poor Quality of Service (QoS) and at the most were useful for home based. Since the late 1980's, vendors have tried to integrate telephony systems with data applications through various computer-telephony integration solutions. In the past few years we have seen tremendous improvements from key market vendors in the new generation of converged applications. One thing is for certain, we are on our way to convergence. In an article from Gartner, AOL Time Warner, Chairman of the Board, Steve Case stated, "Every decade has some word associated with it. In the '80s, it was the PC. In the '90s, it was the Internet. For the rest of this decade, the key word is going to be convergence."

LIST OF ACRONYMS

3GPP	3rd Generation Partnership Program
AAL	ATM Adaptation Layer
ABR	available bit rate
ACF	Admissions Confirm
ADPCM	Adaptive Differential Pulse Code Modulation
AES	Advanced Encryption Standard
AF	Assured Forwarding
AFI	Authority and Format Identifier
AIN	advanced intelligent network
AP	Access Point
APP	Application defined
ARQ	Admissions Request
ATM	asynchronous transfer mode
AVD	alternate voice-data
AVP	Audio Video Profile
BBE	Better than BE
BE	Best Effort
BISDN	Broadband ISDN
BOC	Bell Operating Companies
BRI	Basic Rate Interface
BRQ	Bandwidth Change Request
BSS	Base Station Systems
CAC	Call Admission Control
CAS	channel associated signaling
CBQ	Class-Based Queuing
CBR	constant bit rate
CBWFQ	Class Based Weighted Fair Queuing
CCB	Customer Care and Billing
CCK	Complementary Code Keying,
CCS	common channel signaling
CDV	Cell Delay Variation
CELP	Code Excited Linear Prediction
CES	Circuit Emulation Services
CIR	Committed Information Rate
CLEC	competitive local exchange carriers
CLI	call-level interface

CO	central office
codecs	Coder/Decoders
COPS	Common Open Policy Service
CoS	Class of Service
CPE	customer premise equipment
CPS	Common Part Sub-layer
CRM	Customer Relationship Management
cRTP	RTP header compression
cRTP	Compressed Real Time Protocol
CTI	computer telephony integration
DB-CES	Dynamic Bandwidth CES
DCC	data country code
DDS	Digital Data Service
DE	Discard Eligibility
DES	Data Encryption Standard
DHCP	dynamic host configuration protocol
DID	direct inward dial
Diff-Serv	Differentiated Services
DLSR	Delay since last SR
DS0	digital signal level 0
DSCP	Differentiated Services Code Point
DSP	Digital Signal Processor
DSSS	Digital Spread Spectrum Service
EAP	Extensible Authentication Protocol
EDGE	Enhanced Data for GSM Evolution
EF	Expedited Forwarding
EJB	Enterprise JavaBeans
FDM	frequency division multiplexing
FRAD	frame relay access device
FRF	Frame Relay Fragmentation
FXO	foreign exchange office
FXS	foreign exchange service
GCF	Gatekeeper Confirmation
GPRS	General Packet Radio Service
GPS	Global Positioning System
GRJ	Gatekeeper Reject
GRQ	Gatekeeper Request
HSS	Hughes Software Systems
HTML	hypertext markup language
ICD	international code designator
ICV	integrity check value
IDI	initial domain identifier
IEC or IXC	Interexchange Carrier
IETF	Internet Engineering Task Force
IMTC	International Multimedia Teleconferencing Consortium

IN	intelligent network
IntServ.	integrated services
IP	Internet Protocol
IPDC	Internet Protocol Device Control
IPX	Internet Packet Exchange
ISDN	Integrated Services Digital Network
ISP	Internet service provider
ISUP	ISDN User Part
ITSP	Internet telephony service providers
ITU	International Telecommunications Union
IV	initialization vector
IVR	interactive voice response
IWF	interworking function
JSP	Java Server Pages
KTS	Key Telephone Systems
LAN	local area network
LATAs	Local Access and Transport Areas
LCF	Location Confirmation
LDAP	Lightweight Directory Access Protocol
LD-CELP	Low Delay CELP
LDN	listed directory number
LEC	Local Exchange Carrier
LFI	Link Fragmenting and Interleaving
LRJ	Location Reject
LRQ	Location request
M3UA	MTP3 User Adaptation Layer
MAC	Medium Access Layer
MC	multipoint controller
MCU	Multipoint Control Units
MG	Media Gateways
MGC	Media Gateway Controllers
MGCP	Media Gateway Control Protocol
MIB	Management Information Base
MIPS	millions of instructions per second
MMN	Managed Multiservice Networking
MMS	Multimedia Messaging Service
MMUSIC	Multiparty Multimedia Session Control
MOS	Mean Opinion Score
MP	multipoint processor
MPLS	Multi Protocol Label Switching
MPP	Multilink PPP
MTP	Message Transfer Part
MTU	Maximum Transmission Unit
NAT	Network Address Translation
NIC	Network Interface Card

NIST	National Institutes of Standards and Technology
NMS	Network Management Systems
ODBC	Open Database Connectivity
ODS	operational data stores
OFDM	orthogonal frequency division multiplexing
PBNM	Policy based Network management
PBX	Private Branch Exchange
PCM	pulse code modulation
PCR	Peak Cell Rate
PCT	platform configuration tool
PD	Powered Device
PDP	policy decision point
PDU	protocol data units
PEPs	policy enforcement points
PHB	Per-Hop Behavior
PIBs	Policy Information Bases
PNNI	private network-to-network interface
PoE	Power over Ethernet
PoP	point of presence
POTS	plain old telephone system
PPP	point-to-point
PQ	Priority Queuing
PRI	Primary Rate Interface
PSE	power sourcing equipment
PSTN	Public Switched Telephone Network
pvc	Permanent Virtual Circuits
QoE	Quality of Experience
QoS	quality of service
Q-SIG	Q-Signaling
RAP	Resource Allocation Protocol
RAS	Registration Admission Status
RCF	Registration Confirmation
RF	radio frequency
RRJ	Registration Reject
RRQ	Registration Request
RS	redirect servers
RSVP	Resource Reservation Protocol
RTP	Real Time Transport Protocol
RTS / CTS	request-to-send / clear-to-send
RTSP	Real-Time Streaming Protocol
SAR	Segmentation And Reassembling
SCCP	Signaling Connection Control Point
SCN	Switched Circuit Network
SCTP	Stream Control Transmission Protocol
SDES	Source Description

SDP	Session Description Protocol
SG	Signaling Gateway
SGCP	Simple Gateway Control Protocol
SGMP	Simple Gateway Monitoring Protocol
SIGTRAN	Signaling Transport Internet Draft
SIP	Session Initiation Protocol
SLA	service level agreement
SMDR	Station Message Detail Recording
SMI	Structure of Management Information
SMS	short message service
SMSC	short message service center
SNA	Systems Network Architecture
SNMP	Simple Network Management Protocol
SNR	signal-to-noise ratio
SONET	Synchronous Optical Network
SPPI	Structure of Policy Provisioning Information
SP	service provider
SQL	structured query language
SR	sender reports
SRTS	Synchronous Residual Time Stamping
SS7	Signaling System 7
SSCS	Service Specific Convergence Sub-layer
SSID	service set identifier
SSNs	SubSystem Numbers
STPs	signal transfer points
SVC	switched virtual circuit
T/R	Tip/Ring
TAPI	Telephony Applications Programming Interface
TCAP	Transaction Capability Application Part
TCP	Transmission Control Protocol
TDD	telephone device for the deaf
TDM	Time Division Multiplexing
THD	total harmonic distortion
TKIP	temporal key integrity protocol
TOS	Type of Service
UA	User Adaptation
UAC	user agent client
UAS	User Agent Server
UBR	unspecified bit rate
UCF	Unregister Confirmation
UDP	User Datagram Protocol
URI	Uniform Request Identifier
URQ	Unregister Request
USB	Universal Serial Bus
VAD	Voice Activity Detection

VBR	variable bit rate
VC	virtual circuit
VCC	Virtual Circuit Channel
VHE	Virtual Home Environment
VLAN	Virtual LAN's
VoATM	Voice over ATM
VoFR	Voice over Frame Relay
VoIP	voice over IP
VoP	Voice over Packet
VPN	virtual private network
VQT	voice quality testing
VRVS	Virtual Room Video Service
WAN	wide area network
WAP	wireless application platform
WECA	Wireless Ethernet Compatibility Alliance
WFQ	Weighted Fair Queuing
WiFi	wireless fidelity
WLAN	Wireless Local Area Network
WML	wireless markup language
XML	Extensible Markup Language

CHAPTER 1

DIFFERENCES IN DATA AND VOICE TRAFFIC

In the past, voice and data networks had been kept separate. Legacy network technologies simply could not meet the diverse performance requirements of both voice and data. Advances in networking technology, including fast Ethernet, wire-speed switching, and Policy-Based Quality of Service (QoS) management, have made it possible to build converged voice and data networks. Converged networks enable a new generation of integrated voice/data applications. For example, with converged networks, users of web-based e-commerce applications can view product information while talking with customer service agents in a call center, through a single network connection. The focus of most converged network strategies is Voice over IP (VoIP). VoIP refers to the transmission of telephone conversations over a packet-switched IP network. This IP network could be as small as a single subnet, private LAN, or as large as the public Internet. With VoIP on the LAN, telephone conversations are converted to a stream of IP packets and sent over an Ethernet network. This network is usually restricted to a building or campus. As VoIP technology matures, new conversion methods may emerge. Regardless of the method that is used to convert VoIP traffic for LANs, VoIP traffic will always traverse the LAN as a stream of IP packets. One of the key challenges in implementing VoIP is to design and build an IP-based network that meets QoS requirements and is comparable in performance to conventional circuit-switched telephone networks. The high latency forwarding and best-effort delivery provided by traditional software-based routers is generally not acceptable for streaming traffic such as VoIP because it does not provide maximum latency guarantees or minimum bandwidth guarantees. To understand the core architecture of converged network, we must understand the behavior of voice and data traffic and how networks perform on different network traffic.

Packet Switching versus Circuit Switching

There are primarily two types of networks, circuit switched and packet switched. A good example of circuit switching is the telephone system. Circuit switching is a technique in which a system seeks out the physical "copper" path from the caller's telephone to the receiver's telephone. There is a need to set up an end-to-end path before any data can be sent. This is why it can take several seconds between the end of dialing and the start of ringing for international calls. During this interval, the telephone system is actually hunting for a copper path, and the call request signal must propagate all the way to the destination and be acknowledged. No one else can use the physical path at the same time.

On the other hand, packet switching is a type of network where small units of data are routed through a network based on the destination address in each packet. Compared to circuit switching, which statically reserves the required bandwidth in advance, packet switching acquires and releases bandwidth as and when it is needed. This type of communication between sender and receiver is known as connectionless. Circuit switching is completely transparent. The sender and receiver can use any bit rate, format, or framing method they desire. As for packet switching, the carrier needs to determine these basic parameters first before transmission can occur.

The Internet is based largely on packet switching, and the Net is basically a huge connectionless network joined together. By transmitting data in packets, the same data path can be shared among many users in the network.

	Circuit switching	Packet switching
Dedicated "Copper" Path	Yes	No
Bandwidth available	Fixed	Dynamic
Potentially wasted bandwidth	Yes	No
Store-and-forward transmission	No	Yes
Each packet follows the same route	Yes	No
Call setup	Required	Not required
When can congestion occur?	At setup time	On every packet
Charging	Per minute	Per packet

Table 1.1

A comparison table on circuit switching vs packet switching

Circuit-switched networks are based on Time Division Multiplexing (TDM), in which numerous signals are combined for transmission on a single communications line or channel. Once a connection is made, a connection is established and remains throughout the session. On the other hand, in packet-switched networks relatively small units of data called packets are routed through a network based on the destination address contained within each packet. Breaking communication down into packets allows the same data path to be shared among many users in the network.

In converged networking, packet-switching technology is used to carry the information throughout the network. Converged network supports all of the user's traffic types with packet-based protocols such as ATM, Frame Relay, or IP. Of these, the dominant protocol in the access, edge, and core of the network is the Internet Protocol (IP). A managed IP infrastructure for voice, video, and data transmission is a concept that has begun to revolutionize the industry. It leverages the Internet so that enhanced services can be offered at lower costs and it motivates the development of a family of dynamic, next-generation, real-time applications. Voice, video, and data traffic have significantly different characteristics and are difficult to reconcile within a single network. Data traffic tends to be bursty, consuming large volumes of bandwidth for occasional, short intervals, whereas voice traffic is predictable and requires a steady, low-delay, transmission path from end to end. Video is a hybrid of both.

- Most computer networks currently use "packet switching" rather than "circuit switching"

CIRCUIT SWITCHING
- Circuit Switching: local exchanges connect to remote exchanges
- Each caller may make one call at a time
- Line is unusable by others until call terminated

Figure 1.1

Circuit switching.

Example: POTS (plain old telephone system)

PACKET SWITCHING
- Technique used with digital connections to allow multiple calls to exist on same circuit.
- Data broken into small blocks ("packets")

- Packet includes extra information in Header (serial number, destination address, etc.)
- Each packet routed to its destination individually
- Packets re-assembled into original message when they reach destination

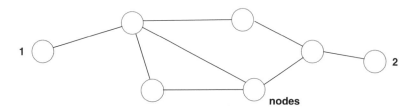

Figure 1.2

Packet switching.

ADVANTAGES OF PACKET SWITCHING
- Network supports many connections simultaneously
- Short messages not delayed by long messages
- More efficient than circuit switching

DISADVANTAGES
- Performance drops when many users share the same network

As far as communication networks are concerned, the notion of switching began with circuit switching. The telephone (or telegraph) company provided an electrical path that allowed my instrument to connect to yours, perhaps with an operator plugging a connector into a jack. The telephone company first combined (or multiplexed) multiple calls on a single physical circuit using Frequency Division Multiplexing (FDM). You can think of these discrete calling paths as the first virtual circuits in the sense that there was no longer a one-to-one ratio between phone calls and wires (or insulators on the utility pole).

Frame Switching

However, FDM turned out to be insufficiently scalable for the demands of telephony. So, in the early 1960s the phone companies began digitizing voice signals and multiplexing them in the time domain using Time Division Multiplexing (TDM).

For example, with TDM a T1 line interleaves 24 phone calls among successive time slots within 1 frame, which consists of 193 bits. Bit 1 through bit 8 are dedicated to channel 1, bit 9 through bit 16 are dedicated to channel 2, and so on, until bit 185 through bit 192 are dedicated to channel 24. The 193rd framing bit is used to synchronize the system. The interleaving process repeats 8000 times each second. (Note that 193 bits per frame times 8000 frames per second equals 1.544Mbits/sec, which is the throughput rate of a T1 line.)

Today, TDM phone call switching is circuit oriented, though the devices that perform the switching function on digital circuits bear little resemblance to the mechanical switches that once selected paths made up of solitary electrical circuits.

One disadvantage to TDM—whether you're making a call with a TDM system or leasing a full-time digital line—is that your cost will be the same whether you fill every time slot with data or transmit nothing (for our purposes, let's consider voice to be just another form of data). As we know, the data transmissions required for many applications are bursty, with intervals of high demand and no demand distributed almost randomly. Thus, TDM-based networks (or any circuit-oriented networks with hard resource allocations) are likely to be inefficient or otherwise not wholly suitable for data traffic.

Ultimately, dedicated circuits are a high-cost form of connectivity, and setting up a circuit via a switching system designed for voice communications results in long circuit-initiation times, as well as high cost.

Packet Switching

In the late 1960s the notion of packet switching was developed. The first commercial outgrowth of this technology was the X.25 network. In spite of outbreaks of Internet fever in many countries, it is still heavily relied on in much of the world. Packets on an X.25 network aren't slotted rigidly in the way that circuits in the TDM system are. Instead, the packets are created and transmitted as needed. Therefore, X.25 service can be priced by the packet or by the byte rather than by connection time or as a full-time circuit; you're using network capacity only when you're sending or receiving data.

Despite this freer form of multiplexing, X.25 networks are still connection oriented, and a session between two nodes still requires a virtual circuit—the virtual circuit has just been unbundled from a fixed time slot.

Each packet in an X.25 network has a Logical Channel Number, or LCN. When a packet comes into a switch, the switch looks up the LCN to decide which port to send the packet out of. The path through the network of packet switches is defined in advance for Permanent Virtual Circuits (PVCs) and established on the fly for Switched Virtual Circuits (SVCs). With SVCs, call setup is required before data transmission can take place. (Incidentally, the protocol data units at Layer 3 are known as packets, while the protocol data units at Layer 2 are called frames, at least when writers are precise. The X.25 protocols include Layer 3 functions, while frame relay, which is essentially X.25 with error correction and flow control removed, remains at Layer 2.)

Connectionless Switching

Local Area Network (LAN), which people began to develop in the 1970s, is a form of connectionless communication. When a Layer 2 bridge connects LANs, a form of frame switching takes place. When a Layer 3 router connects LANs, a form of packet switching takes place. IP and its precursors were the first wide-area connectionless protocols used extensively. Each packet includes its source and destination addresses and moves independently through the network. With these connectionless switching systems, for the first time there was no advance setup of the path needed. There was also no advance agreement on the part of the intermediate stages committing to a specific level of service. There was no state maintained on the network's packet switches, which have no notion of circuits, paths, flows, or any other end-to-end connection.

There's an economic and fault-tolerance factor you should keep in mind when considering the router-using flavor of packet switching. The economic issue revolves around the relative costs of computer cycles and communications circuits. If processing power is cheap, it's not unreasonable to figure out the routes every packet should take in order to make the best use of expensive circuits. If circuits are cheap, you may prefer to set up a mesh of connections using relatively dumb circuit switches, rather than dedicate a bunch of high-powered, special-purpose computers to routing packets.

The fault-tolerance issue (often expressed as the overheated claim that the Internet was designed to survive a nuclear war) is that connectionless networks are more resilient than connection-oriented ones. If a link or a router goes down, the overall system is designed to find routes around the problem. A broken connection-oriented network, designed to remember how to handle each circuit rather than to solve a routing problem for each packet, will likely need manual reconfiguration if it consists of PVCs. Furthermore, if it consists of SVCs, it will at least need to perform a call setup.

While most of the voice links are circuit switched, there does exist some packet switching of voice data in the telephone network. Beyond the local loop, a large part of the signal and control network is packet switched like IP, though the signaling networks generally use a different suite of protocols. Circuit switching and packet switching address the same problem. A lot of data comes into a communications computer (switch) on a set of incoming channels and must be directed to the right outgoing channels. The key difference is how resources are reserved on the channels and within the communications computer.

Where the traditional network deployed today is designed to carry *circuits*—fixed channels across which communications between telephones or other devices are conducted—the telecommunications industry is now preparing for a big technology changeover, what's being called a "next-generation network" based on so-called data packets.

Instead of setting up an explicit channel for each conversation or data flow, all the traffic types—data, video, voice etc.—will travel together across the network links. Each data flow is broken up into packets (variable-length chunks of data), each carrying its own address, with the packets being forwarded by the network to their ultimate destination based on the information they carry with them.

Most importantly, the new network will separate services and infrastructure. In effect, the network is moving from being the equivalent of a rail system, where design fundamentals dictate the specifics of the train (such as the wheel gauge and carriage shape), toward the equivalent of a road system, where any type of vehicle ranging from a motorcycle to a truck can travel on a single, simple infrastructure. Like the road system (and unlike the rail), traffic is granted its own autonomy. Instead of working to a strict, centrally controlled timetable and signaling system, the packets simply set out and steer themselves toward their destinations.

The aptly-named "superhighway" therefore has a series of advantages. It will not only be able to run different sorts of traffic at high speed, but it will be designed in such a way that it can accommodate further change. Unlike the real-world highway, the telecom highway can mandate faster and faster speeds and a vast increase in the number of lanes running between any two points. Like the real-world highway, its strength is its simplicity. The self-addressed packet standard is the equivalent of tarmac. The telecom equivalent of new tires and whole new vehicles can be designed and introduced without detailed reference to the underlying network infrastructure, or roadway. Packet-based networks are not without their problems, however. The main issue for voice traffic is that it has to arrive at its destination not only almost instantaneously but also in coherent chunks to be comprehensible. Connectionless packet routing using the Internet Protocol (IP) is unable to secure guaranteed quality of transmission. Everything depends on how much other traffic there is on the network at any one time. But there are *solutions*. IP is a real break with the past. For the first time the telecom environment will look more like the computer environment, where a standard operating system will enable both equipment vendors and applications providers to innovate to provide value-added services. The result will be good news for users, who will see costs (especially for voice) rapidly diminish. At the same time, the new environment should stimulate a faster innovation cycle with new residential and business services being brought to market in months rather than years or decades.

Data Traffic Characteristics

In the early days of data communications, all networks were circuit-switched, and many still are. More recently, for networks that cover wide areas, the emphasis has shifted to packet switching, simply because it permits the interconnection of far more nodes into a single network. With packet switching, fewer communication channels are required (because channels are shared by many users), and interconnection of networks is much easier to accomplish.

The data layering of communications gives rise to situations where the lower layers of a network are connectionless, but the higher layers establish a connection; the opposite, where the lower layers establish a connection and the upper layers do not, can also occur.

Broadcast Channels

To understand data network we have to know the concept of communication channels of different types of networks. Broadcast channels are, as the name implies, similar to short-wave radio conversations. A node broadcasts on the common channel and all other nodes connected to the channel listen to the message.

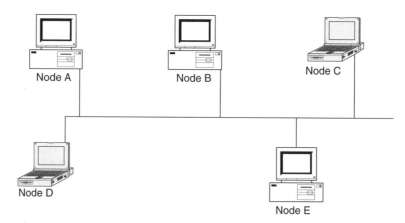

Figure 1.3

Sample of broadcast channel example.

The channel might consist of electromagnetic emissions (that is, radio), or a wire or cable to which all nodes are attached. As with users of CB radio channels, more than one node might begin to transmit at the same time. When that happens, each garbles the other's message. They all must stop and then start over again in such a way that subsequent collisions are not likely to happen.

Alternatively, transmissions can be scheduled so that more than one transmission never takes place at the same time.

Data networks can be classified according to the area over which they extend. A local area network (LAN) consists of a few up to several hundred nodes, but will typically be confined to a few buildings within a few thousand meters of one another. It can consist of subnetworks linked together in certain ways to form the larger, but still local, network. A subnetwork is a portion of a network in which all of the nodes are directly connected; for example, all of the nodes may be connected by one piece of wire.

Metropolitan area networks (MANs) continue to evolve and will be developed primarily by data carriers in response to the demand to interconnect LANs across a metropolitan area. For example, a university might interconnect its campuses. Wide area networks (WANs) are often interconnected LANs or MANs. They can be homogeneous, but are often heterogeneous, that is, interconnecting LANs or MANs that have been built using different technologies. A WAN can span campuses, cities, states, or even continents. Typically, only one node on each LAN or MAN, called a gateway, connects to the WAN. Other nodes communicate with the WAN via the gateway.

Data communications involve the transfer of data between computer programs. Just as humans must share a common language in order to communicate, the programs must have a common protocol. The protocol simply defines the format and meaning of the data that the programs interchange.

An example of a very simple protocol might be the following: Suppose two programs connected by a communications channel need to exchange messages that vary in length. The protocol might specify that the first three characters of each message be numeric characters giving the length of the message itself (in decimal, not counting the first three characters). For the message "HELLO WORLD" (containing 11 bytes including the space), the sending program would transmit the characters "O11HELLO WORLD." The receiving program would accept the first three characters and then, knowing the length of the message, would expect 11 more characters. See Figure 1.4.

Figure 1.4

Computer protocol example.

If the receiving program got a message that had something other than numeric in the first four characters, it would consider that an error. If the sending program stopped sending before all of the characters were received, that would also be considered an error.

Early data communications programs were monolithic—that is, a single large program provided many services and communicated with other programs with a single "low-level" protocol (See Figure 1.5). This did not work well. The programs were hard to change and the protocols would not work well as new technology came along. These problems led to the development of layering concepts.

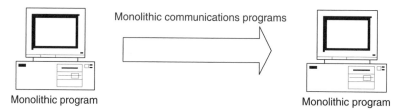

Figure 1.5

Early data communications programs.

A layer is simply a program or set of programs that provides services to the next higher layer and uses services of the next lower layer. A program that resides at the highest layer will typically provide many sophisticated services to the user, but most of these services are actually implemented, directly and indirectly, by the lower layers.

Because a program provides services only to the layer above it and uses services only of the layer below it, a change to any given layer will affect only the layer above it. Layering breaks a single monolithic program into parts that are isolated from one another, making the program easier to write and to change. Layering does, however, extract a performance penalty. There is some overhead associated with moving data through multiple layers.

Protocol Layers

Layering applies to protocols as well as to services. In a system that has a layered architecture, a process communicates only with its peer processes on the same layer; otherwise, as with services, a change to one layer would affect many other layers. Peers communicate with a common protocol, appropriate to the services they provide. Each level may require a protocol that is different. So layers of processes have corresponding layers of protocols.

This concept of layering gives rise, in its simplest form of two layers of services and two layers of protocols, to the service/protocol model shown in Figure 1.6. Learning this basic model is essential to understanding the remainder of the material in this course.

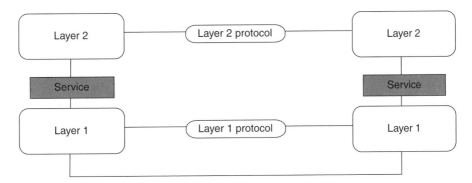

Figure 1.6

Protocol layering.
The diagram illustrates the concept of layering. In this example Layer 1 provides a "service" to the layer above, Layer 2. An example of a service would be transmission of bits across a physical media such as an RS232 cable. Layering breaks a large program into a series of smaller and more manageable tasks. Layering also isolates one layer from another layer.

Layered Communications Systems
The simplest service/protocol model has two layers, but the ideas it illustrates can be extended to an arbitrary number of layers. In fact, real data communications systems typically involve from three to seven layers of services and protocols.

Peer Communications
When two peer programs happen to be running on the same machine, communication will be through the lowest layer in the stack.

Levels of Abstraction
The layers of data communications services can be viewed in a general way. Programs at the lowest layer provide services related to the simplest, most "concrete" form of data: streams of bits. Programs at the highest layer provide services related to the most complex or "abstract" forms of data: data that are ready to be displayed to humans, application program data structures, or, in the case of programs using more advanced object-oriented programming techniques, objects. Programs at the intermediate layers transform the data from the simple to the complex forms, and vice versa. This transformation is sometimes referred to as "abstraction," and you might hear the phrase "levels of abstraction," referring to movement from the concrete to the abstract, that is, from the lowest level to the highest.

Layering and Routing

One reason that protocols vary from layer to layer is because the services offered by each layer to successive layers progress from concrete, bit-oriented services to more abstract, higher-level services for data objects. But protocols differ for another important reason: The lower layers communicate across a single link, while the upper layers must communicate with peers to which the link is indirect, that is, through other nodes.

The bit-stream-oriented protocol at the lowest level supports simple streams of bits flowing between two points. Because the only language understood at this level is "0" and "1," the protocol recognizes only peers to which it is directly connected by the physical link across which the bits flow. It simply has no way of addressing anyone else, and that's not its job anyway.

If communications between peers at all levels were bound by this restriction, communication possibilities would be extremely limited. But a program at a higher level can invoke services from more than one process on the lower level and thus from more than one physical link. Note that the two lowest-level protocols do not need to be identical protocols; in fact, often they will not be. Because higher levels can pass more complex messages, they can include routing information, making it possible for data to flow across more than one communication link, and making large networks possible. So, an important service provided by lower-level protocols in any of the networking systems we will study (which are all layered) is to handle the routing of messages between computers that are not connected by a physical link.

In a system with a layered architecture, programs at a high level deal with more complicated data but don't have to deal with implementation details and routine tasks. For example, in a network the routing of messages is done by lower levels, so upper-level programs need not be aware of it. The data flow only through the highest level that is concerned with handling them. So when a message flows through an intermediate node the data will flow only through the lower-level processes involved with their transmission.

Protocol Stacks

The layering of programs gives rise to layers of protocols, often called "protocol stacks" or "protocol suites," that define a communications standard. Anything or anybody that communicates must share a common protocol with that with which it communicates. Programs that communicate with one another using the same protocol are called peers. Peer programs at each layer provide service to the next higher layer, and use the services of the next lower layer to perform their assigned function. As with services, the lowest-level protocol deals with bits; the highest-level protocol deals with complete data structures or objects that are ready to be used by an application program or displayed to a user.

Data Networks

A data network connects nodes, some of which are hosts to which terminal nodes attach, in two different ways: point-to-point and broadcast. Point-to-point networks fall into two classes: circuit-switched networks, in which a connection is formed between the nodes, as in a telephone network; and packet-switched or connectionless networks, in which packets of data or datagrams are passed from node to node until they reach their destination, like telegrams. All of the nodes in a broadcast network share the same channel, and the network protocol controls access to the channel and avoids or recovers from the collisions that occur when more than one node tries to use the channel at the same time. LANs, local area networks, extend across a single site and consist of one or more subnets, which are usually, but not necessarily, homogeneous. MANs, the metropolitan area networks now evolving, will be developed by data carriers to connect LANs in the same city. WANs are wide area networks, often heterogeneous, which cover many sites, spanning large corporations and sometimes continents. Five common topologies exist for point-to-point networks: star, ring, net, tree, and irregular. Two topologies, which do not use a radio band, have been developed for broadcast: ring, where the ends of the channel are connected, and bus, where they are not.

Voice Traffic Characteristics

To understand voice communications, we have to explore the history of voice networks. When AT&T was divested of its local phone companies on January 1, 1984, two distinct types of telephone companies were legally defined.

1. Local Exchange Carriers (LECs). These consist of the 23 Bell Operating Companies (BOCs) that were created by the divestiture, the former independent telephone companies such as GTE and Contel, and about 1500 small-town telephone companies.
2. Interexchange Carriers (IECs or sometimes IXCs), more often called simply long-distance carriers. These are the former AT&T Long Lines organization and other carriers, such as MCI and Sprint.

The United States was partitioned into Local Access and Transport Areas (LATAs). LATA boundaries conform more or less to the standard metropolitan statistical areas defined by the U.S. Department of Commerce. Originally, they conformed to the boundaries of the areas served by the BOCs.

LECs are prohibited from carrying inter-LATA calls and IECs are prohibited from carrying intra-LATA calls. Even when an LEC covers two LATAs, it must route the call through an IEC. An IEC is represented in a LATA by a Point of Presence (POP). A subscriber can connect to the IEC's facilities only at a POP.

Typically, this connection is made by a line provided by the LEC from the subscriber's premises to the POP (A in the IEC diagram). Alternately, the subscriber can bypass the LEC and connect directly to the POP. For example, the subscriber could install a microwave link to the POP (B).

The providers of inter-LATA services fall into two classes, as follows:

Long-Distance Carriers. Companies that provide long-distance private lines or virtual private networks (switched lines that the subscriber uses in the same manner as dedicated private lines).

Packet Carriers. Companies that provide packet switching services. These companies often use the facilities of the long-distance carriers to construct their networks, of course.

Long-distance carriers compete with offerings in the following specific categories:

1. Voice Grade Service provides leased private analog lines.
2. Digital Data Service (DDS) provides leased digital lines used for data only at speeds ranging from 2400 bps to 19.2 Kbps or, in some cases, 64 Kbps. The DS0 Service provides 64-Kbps digital lines, each providing one DS0 channel. Some carriers provide a clear channel capability, meaning that the carrier does not require any portion of the channel for control signaling. The channel is "clear," and the entire 64 Kbps bandwidth can be used by the subscriber. Otherwise, only 56 Kbps of the 64 Kbps can be used.
3. Fractional T1 lines are the FT1 lines described earlier in this chapter. Some vendors allow the individual FT1 components to be accessed as if they were private voice grade lines or DDS lines while others provide only T1 access.
4. T1 Service provides lines that support voice, data, or video at DS1 speed. Service may be channelized or unchannelized. Channelized means that it is divided into channels by the carrier who then takes care of multiplexing multiple DS0 channels onto it. If unchannelized, the subscriber is responsible for multiplexing.
5. T3 Service is like T1 service but at DS3 speed. It is generally unchannelized, though some vendors do offer channelized service.
6. Some carriers to international locations through certain gateway cities provide international private lines of various types. The subscriber must lease two lines, one to the gateway and one from the gateway to the international location.
7. Switched Data Services are often referred to as software-defined networks or virtual private networks (VPNs). They allow the subscriber to make use of the circuit switching capabilities of the carrier's long-distance facil-

ities to control and monitor their private network. Typically, the carrier provides the subscriber with terminals that interface with the carrier's network control facility. The subscriber can monitor and reconfigure the VPN within certain limits established by the carrier.

With the advent of direct distance dialing (area codes) in the 1960s, it became possible to obtain a high-quality connection to any of over 100 million telephones in the United States by simply dialing the address (phone number) of that device. This was the culmination of 100 years of development of electro-mechanical devices for the routing of calls and of analog devices for the transmission of voices. Voices were transmitted as a continuously varying electrical signal across a pair of wires (a local loop or a subscriber line loop) between the handset of the subscriber (the phone user) and an exchange or end office. (Before digital dialing, end offices had names such as Prospect or Elgin, so telephone numbers began with an abbreviation of the end office name, for example, PR6-6178 or EL3- 1978.) An enterprise could install a Private Branch Exchange (PBX) to switch calls within the enterprise. PBXs were connected to local offices, which connected to trunks.

The electrical signal varied in frequency from about 600 to 3400 Hertz (Hz, cycles per second). Although humans can hear frequencies from about 20 to 20,000 Hz, most speech energy is concentrated in the range of 600–3400 Hz.

Early telephone end offices accomplished switching with human operators, but in the 1930s switching was automated. Before digital dialing, electromechanical rotary switches (in smaller, especially rural, end offices) and crossbar switches connected subscriber lines. Telephones had rotary dials that generated electrical pulses, as many pulses as the number that was dialed (one pulse for "1," two pulses for "2," and so on through ten for "0"). The pulses "stepped" the switches to make the interconnection. Trunk lines connected end offices. As the number of end offices grew, it quickly became necessary to organize the phone system into a hierarchy; too many trunk lines would have been required to interconnect all of the end offices in even a metropolitan area, let alone the whole country (about 20,000 end offices are in use today). By connecting each end office to a toll center and connecting toll centers together with trunk lines, far fewer trunks were required even though any subscriber could reach any other subscriber in the area. Over time, a four-layer hierarchy was developed to interconnect all end offices in the United States.

Long-distance trunk lines were expensive to build, yet even the simple copper wires used for the first long-distance lines were capable of carrying an electrical signal with a much wider bandwidth than was required for voice transmission. A technique called frequency division multiplexing (FDM) was developed that allowed a trunk line with only two pairs of wires to carry many voice conversations (many voice channels) simultaneously. Because of limitations in the technology first developed for FDM, a voice channel was standardized to be 4000 Hz of bandwidth. The additional 600 Hz between 3400 Hz and 4000 Hz was necessary to separate the channels on a multiplexed line so that one channel would not interfere with the next.

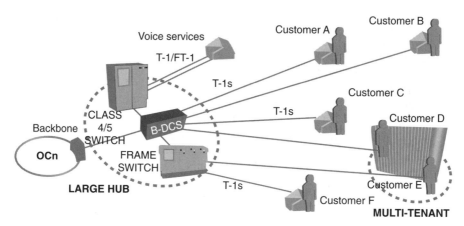

Figure 1.7

Voice communications network.

The Addition of Data

Today's digital telecommunications networks evolved out of the analog switched long-distance networks that were created in the early part of the 20th century. The telephone system was designed for voice communications (See Figure 1.7 Leased Lines diagram). Some errors can be introduced into a voice signal without causing problems for the people at either end of the conversation. For example, a shift in the phase of the signal (essentially a delay in the transmission of the signal) will have little or no effect on voice quality. But such errors can cause problems for data transmission. When the telephone network was an analog network, electromechanical switches could inject a certain amount of noise into circuits. Other instruments in the transmission path, such as multiplexers, could corrupt the signal further. At lower data rates, say 300 or 1200 bps, modems can correctly transmit data with few if any errors even when the line is noisy, but as data rates increase, noise causes more problems. As a practical matter, in the analog network, data could not be transmitted reliably over long distances on switched lines at rates over 4800 bps. Therefore, it became common practice for the telephone companies to "lease" lines to companies for continuous, unswitched use. A company with two computer sites could, for example, lease a line or a set of lines to interconnect the sites.

Leased lines could be "conditioned" by the telephone company to make them usable at higher rates of transmission up to 19.2 bps. That basic limit was imposed by the previously mentioned 4000-Hz bandwidth of voice channels, although recent advances in error correction and data compression techniques have made effective rates significantly beyond that possible. A leased line would usually be composed, not of an actual dedicated set of copper wires, but of a dedicated FDM channel between area offices of the phone company. To obtain effective data transfer rates greater than 19.2 bps, it was then necessary for the subscriber to lease multiple lines and transmit data in parallel.

Leased lines have an additional characteristic that makes them desirable for certain uses: The length of the circuit is known (the telephone company usually is able to provide this information). The length does not usually vary from day to day. Fixed length is important to certain applications that have timing considerations related to actual circuit length, for example, channel extension. (Backup circuits can be leased for an additional charge. The length can vary when the backup is used.)

The data communication terms "half-duplex" and "full-duplex" are closely related to, but not synonymous with, the telephony terms "two-wire" and "four-wire." A half-duplex circuit is one that provides transmission in two directions, but only in one direction at a time. A full-duplex circuit is one that provides transmission in both directions simultaneously. Obviously, and end-to-end four-wire circuit provides full-duplex capabilities. However, two-wire circuits can also be used for full-duplex communication by partitioning the available bandwidth into separate frequency bands for each direction of transmission (derived four-wire). This technique is often utilized when a full-duplex data communications circuit is desired over dial-up (two-wire) facilities. On the other hand, the existence of four-wire circuits does not necessarily imply that full-duplex transmission can be achieved through simultaneous use of both pairs. Very long-distance circuits require echo suppressers that effectively disable one pair of a four-wire circuit while the other pair is in use. Thus only one pair can be used at a time. Long-distance (four-wire) leased lines for data communications typically have the echo suppressers removed so that both pairs can be used simultaneously.

TDM Architecture

Time Division Multiplexing (TDM) replaced FDM. TDM is a scheme in which numerous signals are combined for transmission on a single communications line or channel. Each signal is broken up into many segments, each having a very short duration. TDM multiplexers interleave the output of 12 codecs into a multiplex frame. Three standards prevail for multiplexing—North American, European, and Japanese. All three are based on a DS0 channel that is the pulse code modulation (PCM) output of a codec, and define five or six levels of successive multiplexing. They differ in the number of voice channels multiplexed onto a given level.

The circuit that combines signals at the source (transmitting) end of a communications link is known as a multiplexer. It accepts the input from each individual end user, breaks each signal into segments, and assigns the segments to the composite signal in a rotating, repeating sequence. The composite signal thus contains data from all the end users. At the other end of the long-distance cable, the individual signals are separated out by means of a circuit called a demultiplexer, and routed to the proper end users. A two-way communications circuit requires a multiplexer/demultiplexer at each end of the long-distance, high-bandwidth cable. If many signals must be sent along a single long-distance line, careful engineering is required to ensure that the system will perform properly.

This type of multiplexing combines many digital bit streams with relatively low bit frequency into a single bit stream with a relatively high bit frequency. It is, in essence, a way for many slow communications channels to "time share" a very fast channel. The advantage, of course, is that the cost per bit transmitted on a single fast channel is lower than on slower channels. In TDM, stations are allocated the entire bandwidth of frequencies but only for a small percentage of time (See Figure 1.8). TDM is accomplished by simply interleaving data from

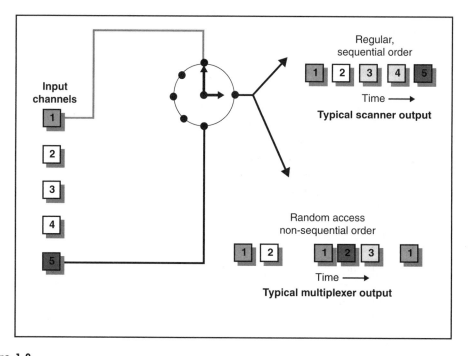

Figure 1.8

Time division multiplexing techniques.

several bit streams. This can be done on a bit basis or on a byte basis (called bit interleaving and byte interleaving). During time interval 1, 8 bits from source channel 1 are transmitted. During successive intervals, bytes from successive source channels are transmitted on the output channel. A complete set of values from each input channel is called a frame.

Convergence Technologies

Telecommunications convergence is the merger of legacy-based time division multiplexing (TDM) architecture with today's packet-switching technology and call-control intelligence, which allows commercial carriers and service providers to consolidate voice and data networks to provide integrated communications services. Convergence technologies are changing the way telecommunications companies will provide voice and data traffic. The public switched telephone network (PSTN) is one of the most reliable communications networks in existence. Using traditional PSTN services as an access point to the Internet has significantly encouraged the growth of data over costly Class-5 facilities, resulting in the need to reengineer this traditional TDM architecture. Convergence technologies will provide a packet-based architecture that combines the speed and efficiency of broadband with the full-featured signaling system 7 (SS7) architecture to create a hybrid network in which carriers and service providers can choose route options based on cost, efficiency, and fault management.

However, convergence is not accomplished by acquiring a single box. Those who intend to compete in the converged market space face significant challenges in choosing the correct equipment and software configured to provide them with the capabilities required to compete in today's Internet economy. Investors, entrepreneurs, and existing vendors must appreciate the complexities of convergence or risk the delays and unnecessary expense of adopting the wrong strategy.

Just as the Internet has predicated this shift to convergence, it has also provided the most compelling means of dominating in the converged market space.

With data and voice networks converging, it is apparent that circuit-switched networks built for voice communications are being adversely affected by the growth in dial data access. Rather than create a new class of equipment to divert data traffic from the PSTN, it is more prudent to develop strategies that leverage existing features found in the voice network today. What makes a converged network work is the ability to control both networks from one platform simultaneously while being able to manage call-control and multiple protocols from different switch applications.

The SS7 network is the vital link between the voice and data networks. SS7 enables routing and control of both voice and data traffic across converged networks while continuing to make intelligent network (IN) services available to the traditional world of telephony. Additionally, SS7 networks connect calls faster and reduce overall operational costs, which stands out as a distinct competitive advantage in a business where every penny counts. For instance, suppose a customer wants to dial a client in another region of the country. According to today's terms, it usually means dialing a 10-digit number that routes out of band (SS7) to its destination. In a converged network, the call would route into a host facility via SS7 A-links, where the host would first check for validation and provide a branded prompt recording, receive the listed directory number (LDN), and route the call via least-cost routing methods to its destination. Based on costs, the call could route out a packet-switched gateway into an ATM cloud to the distance end, where another gateway would receive the call and send it via PSTN to its destination. This routing architecture cuts out the cost of the transcontinental connection.

SS7 has long been a requirement for mainstream carriers but has been financially out of reach for many of the upstarts and growth carriers as a result of the high cost of deployment and licensing. Traditionally, SS7 has been managed by separate SS7 connections at each point of presence (PoP), but this is an expensive solution because of circuits and license fees from each location. A distributed network architecture, a solution that involves running a quad 4-system configuration, offers a cost-saving solution. SS7 servers connect to two disparate signal transfer points (STPs) in different cities for continuous uptime. The SS7 signaling is pulled into the host and distributed using transmission control protocol (TCP)/IP over a clear channel T1 wide-area network (WAN) solution. The key advantage of this distributed SS7 solution lies in the ability to manage multiple sites from a single location in real time. This eliminates the need for multiple licenses for SS7 products and fees for each switch connection in the network.

There are many different applications that handle access to the SS7 network. There are also several issues to consider in choosing a solution. A key preference is for an open standards-based architecture that includes features such as platform compatibility with third-party database applications and a platform configuration tool (PCT) that can be used as a graphical interface to create and manage multiple configurations. The SS7 application should have the capability to be managed remotely through the same interfaces used to communicate with the rest of the platform. SS7 and IP signaling should both be supported, thus enabling these services to run across convergent voice and data networks.

CHAPTER 2

NETWORKS SUPPORTED—KEEPING EXISTING INFRASTRUCTURES IN PLACE

One of the most important phases of a converged network is to choose the right technology vendor or network integrator, one that can provide a seamless solution without forklifting your existing network. A network manager or administrator should consider an open converged platform architecture so that it is easier to deploy converged and next-generation solutions without compromise. Converged communication services drive the current model and pace of business, enabling a whole new way of transacting commerce and interacting with customers and employees. The emergence of global competition and the continued domination of the Internet have driven a need for much more sophisticated multifunctional communication services in order for businesses to reach customers. Carriers are responding to the challenge through the building of much more versatile platforms. This occurs in both new network deployments and the transformation of their traditional singular communication networks such as ATM, TDM, Frame Relay, from IP to Multiservice Network Architectures.

Converged network supports all of the user's traffic types with packet-based protocols such as ATM, Frame Relay, or IP. Of these, the dominant protocol in the access, edge, and core of the network is the Internet Protocol (IP). The concept of a managed IP infrastructure for voice, video, and data transmission has begun to revolutionize the industry. It leverages the Internet so that enhanced services can be offered at lower costs and it is motivating the development of a family of dynamic, next-generation, real-time applications. For traditional carriers, the importance of the co-existence of existing and newer technologies and equipment is an important reality. Next-generation multiservice networks, based on packetized technologies, will not entirely push out existent circuit-switching equipment and solutions for many years. This is where a network element called a "media gateway" is needed—it performs both signaling and media format conversion between different network encoding and transmission standards. For new communications service providers that don't have a legacy of traditional networks, co-existence is moot. Flexible and versatile multiservice network architectures can provide an immediate competitive advantage. Although these are relatively new technologies, there are still issues of proven capability to be achieved. However, these new service providers must still work within existing networks to maximize customer coverage and appeal. The key to success in this new communications market is the ability to rapidly create new communications solutions that offer carriers and other communication services providers a competitive edge in attracting new customers as well as retaining existing ones. The adoption of open standards in equipment development generates a virtuous circle where a market for outsourced components fosters competition, which in turn drives innovation, so visionary network equipment makers are looking to feature and function development partners who adhere to open standards for leading-edge technology development.

Before discussing TDM, ATM, Frame Relay, Internet Protocol (IP) and how a company can migrate VoIP, we need to understand the basic rules of traditional voice communications.

Voice Networking

Basic voice technology has been available for more than 100 years. During that time, the technology has matured to the point at which it has become ubiquitous and largely invisible to most users. This legacy of slow evolution continues to affect today's advanced voice networks in many ways, so it is important to understand the fundamentals of traditional voice technology before emulating it on data networks.

Traditional analog telephone instruments used for plain old telephone service (POTS) use a simple two-wire interface to the network. They rely on an internal two-wire/four-wire hybrid circuit to both transmit and receive signals. This economical approach has been effective but requires special engineering regarding echo.

Basic Telephony

Three types of signaling are required for traditional telephony: supervision, alerting, and addressing. Supervision monitors the state of the instrument—for example, allowing the central office or PBX to know when the receiver has been picked up to make a call or when a call is terminated. Alerting concerns the notification of a user that a call is present (ringing) or simple call progress tones during a call (such as busy, ringback, and so on). Finally, addressing enables the user to dial a specific extension.

In addition to signaling, telephony services also provide secure media transport for the voice itself, analog-to-digital conversion, bonding and grounding for safety, power, and a variety of other functions when needed. Analog voice interfaces have evolved over the years to provide for these basic functions while addressing specific applications. Because basic POTS two-wire analog interfaces operate in a master/slave model, data equipment must emulate two basic types of analog interfaces: the user side and the network side. The user side (telephone) expects to receive power from the network as well as supervision.

A foreign exchange service (FXS) interface is used to connect an analog telephone, fax machine, modem, or any other device that would be connected to a phone line. It outputs 48 vdc power, ringing, and so on, and it accepts dialed digits. The opposite of an FXS interface is a foreign exchange office (FXO) interface. It is used to connect to a switching system providing services and supervision and, at the other end, it expects the switch to provide supervision and other elements. (Why "foreign"? The terms FXS and FXO were originally used within telephone company networks to describe provision of telephone service from a central office other than normally assigned.)

Within FXS and FXO interfaces, it is also necessary to emulate variants in supervision. Typical telephones operate in a loop start mode. The telephone normally presents a high impedance between the two wires. When the receiver goes off-hook, a low-impedance closed circuit is created between the two wires. The switch, sensing current flow, then knows that the receiver is off-hook and applies a dial tone. The switch also checks to be sure that the receiver is on-hook before sending a ringing signal. This system works well for simple telephones, but it can cause problems on trunks between PBXs and COs with high activity. In that situation, the remote end and the CO switch can both try to seize the line at the same time. This situation, called *glare*, can freeze the trunk until one side releases it. The solution is to short tip or ring to ground as a signal for line seizure, rather than looping it. This is called *ground start*.

After the line is seized, it is necessary to dial the number. Normal human fingers cannot outrun the dial receivers in a modern switch, but digits dialed by a PBX can. In that case, many analog trunks use a delay start or wink start method to notify the calling device when the switch is ready to accept digits.

Another analog interface often used for trunking is E&M. This is a four- or six-wire interface that includes separate wires for supervision in addition to the voice pair. *E&M* stands for "ear and mouth" or "Earth and magneto" and is derived from the early telephony days. The E&M leads are used to signal on-hook and off-hook states.

An analog interface works well for basic trunk connections between switches or PBXs, but it is uneconomical when the number of connections exceeds six to eight circuits. At that point, it is usually more efficient to use digital trunks. In North America, the T1 (1.544 Mbps) trunk speed is used, consisting of 24 digitized analog voice conversations. In other parts of the world, E1 (2.048 Mbps) is used to carry 30 voice channels. (Engineers refer to the adoption of E1 and T1 internationally as "the baseball rule"—there is a strong correlation of countries that play baseball to the use of T1. Therefore, the United States, Canada, and Japan have the largest T1 networks, while other countries use E1.)

The first step in conversion to digital is sampling. The Nyquist theorem states that the sampling frequency should be twice the rate of the highest desired frequency. Early telephony engineers decided that a range of 4000 Hertz would be sufficient to capture human voices (which matches the performance of long analog loops). Therefore, voice channels are sampled at a rate of 8000 times per second, or once every 125 ms. Each one of these samples consists of an 8-bit measurement for a total of 64,000 bits per second to be transmitted. As a final step, companding is used to provide greater accuracy of low-amplitude components. In North America, this is μ-law (mu-law), while elsewhere it is typically A-law. For international interworking purposes, it is agreed that the North American side will make the conversion.

To construct a T1, 24 channels are assembled for a total of 1.536 Mbps, and an additional 8 bits are added every 125 ms for framing, resulting in a rate of 1.544 Mbps. Often, T1 frames are combined into larger structures called SuperFrames (12 frames) and Extended-SuperFrames (24 frames). "Robbing bits" from the interior frames can then transmit additional signaling.

Basic T1 and E1 interfaces emulate a collection of analog voice trunks and use robbed bit signaling to transfer supervisory information similar to the E&M analog model. As such, each channel carries its own signaling, and the interface is called channel-associated signaling (CAS). A more efficient method uses a common signaling channel for all the voice channels. Primary Rate Interface for ISDN is the most typical example of this common channel signaling (CCS). If voice/data integration is to be successful, all of these voice interfaces must be supported to provide the widest possible range of applications. Over the years, users have grown to expect a certain level of performance, reliability, and behavior of a telecommunications system, which must be supported while going forward. All these issues have been solved by various packet voice systems today so that users can enjoy the same level of support to which they have become accustomed. Before we jump into ATM, Frame Relay, and IP, we need to step back and take a look at Time Division Multiplexing (TDM).

Time Division Multiplexing (TDM)

TDM (time division multiplexing) is a scheme in which numerous signals are combined for transmission on a single communications line or channel. Each signal is broken up into many segments, each having a very short duration.

The circuit that combines signals at the source (transmitting) end of a communications link is known as a multiplexer. It accepts the input from each individual end user, breaks each signal into segments, and assigns the segments to the composite signal in a rotating, repeating sequence. The composite signal thus contains data from all the end users. At the other end of the long-distance cable, the individual signals are separated out by means of a circuit called a demultiplexer and then routed to the proper end users. A two-way communications circuit requires a multiplexer/demultiplexer at each end of the long-distance, high-bandwidth cable.

If many signals must be sent along a single long-distance line, careful engineering is required to ensure that the system will perform properly. An asset of TDM is its flexibility. The scheme allows for variation in the number of signals being sent along the line, and constantly adjusts the time intervals to make optimum use of the available bandwidth. The Internet is a classic example of a communications network in which the volume of traffic can change drastically from hour to hour. In some systems, a different scheme, known as frequency division multiplexing (FDM), is preferred.

Voice over ATM

Many companies have ATM backbone infrastructure and deploying convergence within their network and outside their network. It is crucial to understand how an ATM network is configured before even considering convergence for voice/video/data. ATM (asynchronous transfer mode) is a dedicated-connection switching technology that organizes digital data into 53-byte cell units and transmits them over a physical medium using digital signal technology. Individually, a cell is processed asynchronously relative to other related cells and is queued before being multiplexed over the transmission path. Because ATM is designed to be easily implemented by hardware (rather than software), faster processing and switch speeds are possible. The pre-specified bit rates are either 155.520 Mbps or 622.080 Mbps. Speeds on ATM networks can reach 10 Gbps. Along with Synchronous Optical Network (SONET) and several other technologies, ATM is a key component of broadband ISDN (BISDN).

From the start, ATM was designed to be a multimedia, multiservice technology. It has been accepted by the marketplace for its ability to deliver high-speed data services. Until the recent past, its potential for deploying for voice services was overlooked. With the competitiveness of today's market, the network operators and the service providers have been continuously striving to reduce operating costs and lift network efficiency. They recognized that significant economic benefits could be achieved once the data traffic and voice traffic were integrated onto a single network. Because ATM has been around for a decade or more, claiming to be a multimedia technology, most of the service providers have started installing single ATM infrastructure to support voice, video, and data transfer. Initially, there were a lot of technical issues that were unaddressed, which basically hampered the growth of VTOA. Thanks to the efforts of the ATM Forum and its members, these issues have been addressed and it is now possible to build and operate an ATM network to meet the needs of various types of voice application.

Many in the telecommunications industry believe ATM will revolutionize the way networks are designed and managed because ATM combines the best features of two common transmission methods. Its connection-oriented nature makes ATM a reliable service for delay-sensitive applications such as voice, video, or multimedia. Its pliant and efficient packet switching provides flexible transfer of other forms of data. In a relatively short period of time, ATM has gained a worldwide reputation as the ultimate means of solving end-to-end networking problems. The popularity of ATM has grown such that virtually every LAN equipment vendor and service provider is racing to develop ATM-based products.

Voice over ATM (VoATM) can be supported as standard pulse code modulated (PCM) voice via circuit emulation (AAL1, described later) or as variable bit rate voice in ATM cells as AAL2 (also described later). ATM offers many advantages for transport and switching of voice. First, quality of service (QoS) guarantees can be specified by service provisioning or on a per-call basis. In addition, call setup signaling for ATM switched virtual circuits (SVCs), Q.2931, is based on call setup signaling for voice ISDN. Administration is similar to circuit-based voice networks.

However, VoATM suffers from the burden of additional complexity, and incomplete support and interoperability among vendors. It also tends to be more expensive because it is oriented toward all-optical networks. Most importantly, ATM is typically deployed as a WAN Layer 2 protocol and therefore does not extend all the way to the desktop. Nevertheless, ATM is quite effective for providing trunking and tandem switching services between existing voice switches and PBXs.

Voice over Frame Relay (VoFR) has become widely deployed across many networks. Like VoATM, it is typically employed as a tie trunk or tandem-switching function between remote PBXs. It benefits from much simpler administration and relatively lower costs than VoATM, especially when deployed over a private WAN network. It also scales more economically than VoATM, supporting links from T1 down to 56 kbps. When deployed over a carefully engineered Frame Relay network, VoFR works very well and provides good quality. However, voice quality over Frame Relay can suffer depending on network latency and jitter. Although minimal bandwidth and burstiness are routinely contracted, latency and jitter are often not included in service level agreements (SLAs) with service providers. As a result, voice performance can vary. Even if quality is good at first, voice quality can degrade over time as a service provider's network becomes saturated with more traffic. For this reason, many large enterprise customers are beginning to specify latency and jitter, as well as overall packet throughput from carriers. In these situations, Voice over Frame Relay can provide excellent service.

Voice over IP (VoIP) has begun to be deployed in recent years as well. Unlike Voice over Frame Relay and Voice over ATM, Voice over IP is a Layer 3 solution. It offers much more value and utility because IP goes all the way to the desktop. This means that in addition to providing basic tie trunk and tandem-switching functions to PBXs, VoIP can actually begin to replace those PBXs as an application. As a Layer 3 solution, VoIP is routable and can be carried transparently over any type of network infrastructure. This includes both Frame Relay and ATM. Of all the packet voice technologies, VoIP has perhaps the most difficult time supporting voice quality because Quality of Service (QoS) cannot be guaranteed. Normal applications such as TCP running on IP are insensitive to latency but must retransmit lost packets due to collisions or congestion. Voice is much more sensitive to packet delay than packet loss. In addition to normal traffic congestion, QoS for VoIP is often dependent on lower layers that are ignorant to the voice traffic mingled with the data traffic.

The ATM Forum and the ITU have specified different classes of services to represent different possible traffic types for VoATM. Designed primarily for voice communications, constant bit rate (CBR) and variable bit rate (VBR) classes have provisions for passing real-time traffic and are suitable for guaranteeing a certain level of service. CBR, in particular, allows the amount of bandwidth, end-to-end delay and delay variation to be specified during the call setup. Designed principally for bursty traffic, unspecified bit rate (UBR) and available bit rate (ABR) are more suitable for data applications. UBR, in particular, makes no guarantees about the delivery of the data traffic.

The method of transporting voice channels through an ATM network depends on the nature of the traffic. Different ATM adaptation types have been developed for these different traffic types, each with its own benefits and detriments. ATM adaptation layer 1 (AAL1) is the most common adaptation layer used with CBR services.

Unstructured AAL1 takes a continuous bit stream and places it within ATM cells. This is a common method of supporting a full E1 byte stream from end to end. The problem with this approach is that a full E1 may be sent, regardless of the actual number of voice channels in use. (An EI is a wide-area digital transmission scheme used predominantly in Europe, which carries data at a rate of 2.048 Mbps.)

Structured AAL1 contains a pointer in the payload that allows the digital signal level 0 (DS0) structure to be maintained in subsequent cells. This allows network efficiencies to be gained by not using bandwidth for unused DS0s. (A DS0 is a framing specification used in transmitting digital signals over a single channel at 64 kbps on a T1 facility.)

The remapping option allows the ATM network to terminate structured AAL1 cells and remap DS0s to the proper destinations. This eliminates the need for permanent virtual circuits (PVCs) between every possible source/destination combination. The major difference from the previous approach is that a PVC is not built across the network from edge to edge.

VoATM Signaling

Figure 2.1 describes the transport method in which voice signaling is carried through the network transparently. PVCs are created for both signaling and voice transport. First, a signaling message is carried transparently over the signaling PVC from end station to end station. Second, coordination between the end systems allows the selection of a PVC to carry the voice communication between end stations.

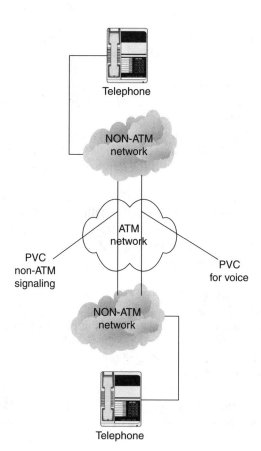

Figure 2.1

The VoATM signaling transport model describes the transport method, in which voice signaling is carried through the network transparently.

At no time is the ATM network participating in the interpretation of the signaling that takes place between end stations. However, as a value-added feature, some products are capable of understanding channel associated signaling (CAS) and can prevent the sending of empty voice cells when the end stations are on-hook.

VoATM Addressing

ATM standards support both private and public addressing schemes. Both schemes involve addresses that are 20 bytes in length (shown in Figure 2.2).

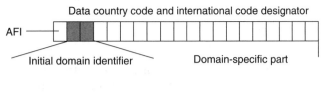

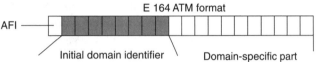

Figure 2.2

ATM supports a 20-byte addressing format.

The *Authority and Format Identifier* (*AFI*) identifies the particular addressing format employed. Three identifiers are currently specified: data country code (DCC), international code designator (ICD), and E.164. A standard body administers each one. The second part of the address is the initial domain identifier (IDI). This address uniquely identifies the customer's network. The E.164 scheme has a longer IDI that corresponds to the 15-digit ISDN network number. The final portion, the domain-specific part (DSP), identifies logical groupings and ATM end stations.

In a transport model, you don't need to be aware of the underlying addressing used by the voice network. However, in the translate model, the capability to communicate from a non-ATM network device to an ATM network device implies a level of address mapping. Fortunately, ATM supports the E.164 addressing scheme that is employed by telephone networks throughout the world.

VoATM Routing

ATM uses a *private network-to-network interface* (*PNNI*), a hierarchical link-state routing protocol that is scalable for global usage. In addition to determining reachability and routing within an ATM network, it is also capable of call setup.

A virtual circuit (VC) call request causes a connection with certain QoS requirements to be requested through the ATM network. The route through the network is determined by the source ATM switch. This is based on what it determines is the best path through the network and on the PNNI protocol and the QoS request. Each switch along the path is checked to determine whether it has the appropriate resources for the connection.

When the connection is established, voice traffic flows between end stations as if a leased line existed between the two. This specification spells out routing in private networks. Within carrier networks, the switch-to-switch protocol is B-ICI. Current research and development of integrated non-ATM and ATM routing will yield new capabilities to build translate-level voice and ATM networks.

VoATM and Delay

ATM has several mechanisms for controlling delay and delay variation. The QoS capabilities of ATM allow the specific request of constant bit rate traffic with bandwidth and delay variation guarantees. The use of VC queues allows each traffic stream to be treated uniquely. Priority can be given for the transmission of voice traffic. The use of small, fixed-size cells reduces queuing delay and the delay variation associated with variable-sized packets.

Benefit of Voice over ATM

With the standards now in place, it is possible for packet switching techniques such as Frame Relay or ATM to deliver high-quality speech. Some of the intrinsic advantages ATM has over other network technologies are listed here.

- The very format of the cell was arrived at by considering data, voice, and video payload requirements. ATM cells are of fixed size, 53 bytes each with 48 bytes for payload and 5 for ATM header. This helps in reducing the packetization delay significantly, which is one of the major delay parameters.
- It supports extensive QoS (Quality of Service), which allows voice traffic to be transmitted across the network in a reliable jitter-free way.
- Various ATM Adaptation Layers (AALs) support various service classes capabilities.
- ATM switches have always been designed with effective traffic management capabilities—for example, call admission control, usage parameter control, traffic shaping, etc.
- It allows a single network for voice, data, and video.
- Interworking with PSTN is relatively straightforward.

Though ATM is equipped with transferring voice over the network efficiently, ATM onto the desktop would not be popular enough until the turn of the millennium. The reasons being, there are enough competing technologies, like 100-Mbps Ethernet/Gigabit Ethernet that provide similar services with minimal infrastructure upgrade to be done. But, when WAN is considered, ATM has its own niche over its competing technologies for the same reasons previously discussed. While designing and engineering voice over an ATM WAN, there are sets of design issues that need to be addressed. Some of them are as follows.

Technical Challenges

A packetized approach to transmit voice faces a number of technical challenges, which spring from the real-time or interactive nature of the voice traffic. Some of the challenges that need to be addressed are discussed next.

Echo

This is a phenomenon where the transmitted voice signal gets reflected back. It is due to the unavoidable impedance mismatch and the four-wire/two-wire conversion between the telephone handset and the communication network. It can, depending on the severity, disrupt the normal flow of conversation. Its severity depends on the round-trip time delay. It is found that if the round-trip time delay is more than 30 ms, the echo becomes significant, making it difficult to have a normal conversation.

End-to-End Delay

Voice is most sensitive to delay and mildly sensitive to variations in delay (jitter). It is highly critical that the delay is kept at a bare minimum to hold an interactive communication end-to-end. It has been found that delay can have two effects on communication performance. Delay can interfere with the dynamics of voice communication, in the absence of noticeable echo, whereas in the presence of noticeable echo, increasing delay makes echo effects worse. When the delay reaches above 30 ms, echo cancellor circuits are required to control the echo. Once the echo cancellor circuits are in place, network delays can be allowed to reach up to 150 ms without further degrading the voice quality.

According to the ITU-T Recommendation G.114, Table 2.1 shows allowable delay limits for one-way transmission time for connections with adequate echo control.

Delay	Acceptability
0–150 ms	acceptable to most user applications
150–400 ms	acceptable when the impact on user's application is minimal
more than 400 ms	unacceptable

Table 2.1

Delay limits

Delay occurs in ATM networks because of one or more of the following reasons:

a)   Packetization Delay (or cell construction delay)
This is the time taken to fill in a complete packet/cell before it is transmitted. Normal PCM (Pulse Code Modulation) encoded voice samples arrive at the rate of 64 Kbps, which means it takes around 6 ms to fill the entire 48-byte payload of the ATM cell. The problem can be addressed either with partially filled cells or by mulitplexing several voice calls into a single ATM VCC (Virtual Circuit Channel).

b)   Buffering Delay
Sometimes, due to the delay in transit, some cells might arrive late. If this happens, SAR (Segmentation And Reassembling) function provided by the Adaptation layer might have to under-run with no voice data to process, which would result in gaps in the conversation. To prevent this the receiving SAR function would accumulate a buffer of information before starting the reconstruction. In order to ensure no under-runs occur, the buffer size should be kept in such a way it exceeds the maximum predicted delay. The size of the buffer translates into delay, as each cell must progress through the buffer on arrival at the emulated circuit's line rate. This implies that the Cell Delay Variation (CDV) has to be controlled within the ATM network.

c)   Encoding Delay
This is the processing time taken by the compression algorithms to encode the analog signal to digital form.

Silence Suppression

Voice is inherently variable. It is found that on an average, human voice has a speech activity factor of about 42 percent. There are pauses between sentences and words with no speech in either direction. Also, voice communication is a half-duplex, that is, one person is silent while the other speaks. One can take advantage of these two characteristics to save bandwidth by halting the transmission of cells during these silent periods. This is known as silence suppression.

Compression Algorithms

ADPCM (Adaptive Differential Pulse Code Modulation) and CELP (Code Excited Linear Prediction) are the two major compression algorithms that are used. Now, LD-CELP (Low Delay CELP), a derivative of CELP, is the compression algorithm most commonly used on any voice signal. This has been standardized by ITU as ITU G.728. This provides a toll quality voice at 16 Kbps with low encoding/decoding delay. Table 2.2 compares the various compression techniques for their bandwidth usage, MIPS (millions of instructions per second), and delay.

Algorithm	Bandwidth	MIPS (C5x DSP)	Total Codec Delay (msec)	Application
PCM	64 kbit/s	n/a	0.25	PSTN
ADPCM (G.726)	32 kbit/s	10	0.25	PSTN, cordless phones
CS-ACELP (G.729)	8 kbit/s	30	25	VoFR, VoATM, VoIP
CS-ACELP Annex A (G.729A)	8 kbit/s	20	25	VoFR, VoATM, VoIP
LD-CELP (G.728)	16 kbit/s	40	1.25	PSTN
MP-MLQ (G.723.1)	5.3/6.3 kbit/s	30	67.5	Multimedia and VoIP

Table 2.2

Comparison between various compression algorithms

Signaling

This relates to the efficient utilization of resources and the transfer of control and signaling information. There are two parts in a voice call—the actual voice samples and the signaling information, such as dialed number, the on-hook/off-hook status of the call, and other routing and control information. This signaling can be encoded and may be sent as Common Channel Signaling (CCS), where signaling information from different channels is aggregated into a single signaling channel, and Channel Associated Signaling (CAS), where signaling information is embedded within each discrete voice channel.

Synchronization

The transport of voice demands that the data be synchronized between the speaker and the listener. There are two standard mechanisms that are used to achieve synchronization between point-to-point applications. They are Adaptive Clocking and Synchronous Residual Time Stamping (SRTS). These mechanisms work by adjusting the clock rate at one end of the circuit based on the clock rate of the other end.

The above-mentioned mechanisms work effectively only in the master–slave environment or point-to-point communication. When multipoint services are in operation, it is not possible for a slave to adjust its clock based on two or more difference signals coming from different master sites. For multipoint service, it is easy to adopt an externally synchronized model where each node in the network is synchronized to some external clock source.

Standards and Specifications of ATM

Various applications are available for the transport of voice over an ATM network. Each application has differing requirements for voice transport based on what class of network operators they are defined in. National or International Operators typically have an extensive PSTN service operating over SDH/SONET or PDH infrastructure. When bandwidth is limited, there will be a requirement to integrate voice and data traffic, for reasons of efficiency, into a single ATM network. Within the local loop, ATM may be a valuable solution for the carriage of voice and data to business premises.

Alternate Carriers or Value-Added Network Suppliers take up licenses to provide communication services in competition with the incumbent national operators. Instead of having their own transmission infrastructure, they buy bandwidth from the primary operator. Cost and limited availability of bandwidth demand ATM efficiency as well as the integration of voice and data services. An example of alternate carriers is cellular phone operators.

Private/Enterprise networks buy bandwidth at commercial (retail) rates and achieve the most they can with the resources on hand. Such organizations will have already deployed a TDM network utilizing E1 or T1 links. They will be looking to integrate these solutions into a new ATM network and gain improvements in network performance and efficiency by moving from TDM to statistical multiplexing. With this scenario, two voice transport models have come up. One is known as "voice trunking" and the other as "voice switching."

a) Voice Trunking

This involves tunneling of voice traffic over the ATM network between two fixed end points. This is an appropriate mechanism for connecting of voice switch sites, PBXs, or message switching centers. Here, network is not needed to process or terminate signaling, other than the opportunity to use the signaling to detect idle channels.

b) Voice Switching

Here, the ATM network interprets the voice signaling information and routes the call across the network. In voice switching, the ATM switch receives a call and routes it to the appropriate destination. The VPN network fits appropriately for this kind of functionality. This type of network solution requires that ATM networks interpret the signaling provided from the voice network. Previously, this posed a major challenge as the signaling standards were proprietary. At the present, many vendors provide ATM-based solutions that are able to interpret the signaling provided by their own voice switches. Widespread adoption of ISDN and QSIG voice signaling standards are allowing ATM vendors to offer a standardized voice signaling interface.

From the foregoing analysis of various network operators and transport models, one could see a common set of network requirements emerge. These demonstrate that the minimum characteristics shown in Table 2.3 have to be supported to implement voice trunking.

Characteristics	Necessity
Adaptation	A mechanism to encode voice samples into ATM while meeting the delay and real-time constraints of voice traffic
Signaling	A mechanism that allows the end-to-end transport of voice signaling (Common Channel or Channel Associated) with the voice traffic
Low cross network delay (Latency)	To minimize delay issues and allow normal interactive conversation (This is not a requirement for broadcast applications)
Limited variation in delay	To minimize delays and allow effective echo cancellation

Table 2.3

Objectives to be achieved—1

Apart from these basic demands, to support a complex voice-switched solution or implement efficient statistical mulitplexing, the requirements shown in Table 2.4 also have to be met.

Characteristics	Necessity
Signaling Analysis	To allow set up and tear-down of circuits on demand (or allocation and release of resources)
Call switching and routing mechanisms	To allow configuration of "real-world" VPN applications
Silence suppression or Variable Bit Rate (VBR) encoding	To realize statistical gain (provides at least a doubling in performance)
Call Admission Control (CAC)	To ensure quality of service is preserved
Network resource utilization	To allow statistical overbooking of network resources

Table 2.4

Objectives to be achieved—2

All the proposed ATM solutions will be measured against the current generation of TDM solutions that are successfully deployed. ATM Forum's VTOA workgroup, which is actively involved in developing standards for voice over ATM networks, has developed a set of solutions or standards that could offer direct commercial or operational benefits to any user. These standards allow voice traffic to be carried over an ATM network more efficiently than any packet- or TDM-based infrastructures.

The ATM Forum has defined three principal approaches to carrying voice over an ATM network. These are

- Circuit Emulation Services (CES), which are used to carry full or fractional E1/T1 circuits between end points.
- Dynamic Bandwidth CES (DB-CES).
- ATM Trunking of Narrowband Services using AAL2.

Circuit Emulation Services (CES)

Circuit Emulation allows the user to establish an AAL1 connection to support a circuit, such as a full T1 or E1, over the ATM backbone. In using CES, the ATM network provides a transparent transport mechanism for various CBR (Constant Bit Rate) services based on AAL1. It specifically covers the following types of CBR services:

1. Structured DS1/E1 n * 64 kbps (fractional DS1/E1) service.
2. Unstructured DS1/E1 (1.544 Mbps, 2048 Mbps) service.
3. Unstructured DS3/E3 (44,736 Mbps, 34,368 Mbps) service.

4.   Structured J2 n * 64 kbps (fractional J2) service.
5.   Unstructured J2 (6312 Mbps) service.

Figure 2.3 shows two ATM circuit emulation services (CES) interworking functions (IWFs) connected to an ATM network via physical UNI interfaces. The other sides of the CES-IWFs are connected to standard CBR circuits (e.g., DS1/DS3, J2, or E1/E3), which may originate, for example, on a user's PBX. The job of the two IWFs is to extend this CBR circuit over the ATM network. This means an ATM portion of the connection should retain the bit integrity— that is, analog signal loss cannot be inserted and voice echo control cannot be performed. These must be performed either by the DTE or before the ATM CES IWF is encountered. An ATM UNI physical interface has two characteristics that are relevant when supporting CES.

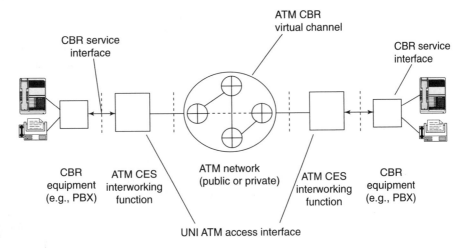

Figure 2.3

ATM circuit emulation.

1.   Bandwidth—The ATM interface must provide adequate bandwidth to carry n * 64 or unstructured traffic after segmentation.
2.   Timing—The ATM interface can be used to convey timing traceable to a primary reference source from the ATM network to the CES interworking function, where the external connection to network timing is not supported.

An ATM network providing CES should also consider QoS parameters such as Peak Cell Rate (PCR) and Cell Delay Variation (CDV). There are standards that specify what values are optimum for these parameters.

CES's advantage is the simplicity of implementation. The ATM network is used to provide virtual replacements for physical links in an existing network. Still,

CES has two limitations. First, it is unable to provide any statistical multiplexing. It does not differentiate between idle and active timeslots. This means all idle traffic is carried. Therefore, CES voice transport consumes 10 percent more bandwidth than would be required to transfer the same voice traffic over leased circuits. Second, it is often implemented as a point-to-point service, providing the transport of the contents of one network physical interface to another physical network interface. This can prevent the implementation of some network topologies, and can result in increased network cost.

Dynamic Bandwidth CES (DBCES)

The limitations in CES resulted in the development of a new standard from the VTOA workgroup of the ATM Forum, which is referred to as "Dynamic Bandwidth Circuit Emulation Services DB-CES." The objective of this standard is to detect active or inactive timeslots of a legacy TDM trunk from a PBX or multiplexer so that inactive timeslots can be dropped from the next ATM structure to allow you to use and reutilize this bandwidth for other services such as CBR, VBR, UBR, and ABR applications.

ATM Trunking of Narrowband Services Using AAL2

The foregoing CES mechanisms treat voice as being a constant stream of information encoded as a CBR stream. But in the real sense, voice is a combination of talk spurts and silence. So bandwidth is wasted when transmitting the silence. These mechanisms also typically minimize the problems of cell construction delay by transmitting the voice as an uncompressed 64 Kbps. Such approaches deny the network operator a chance to gain significant bandwidth from voice compression technologies. To address these limitations, the ATM Forum came up with the advanced mechanism for the transport of voice as a variable bit rate (VBR) compressed stream.

AAL2 is a new ATM Adaptation Layer, specified in ITU-T Recommendation I.363.2 (1997) with the specific mandate to provide efficient Voice-Over-ATM services. AAL2 supports, in addition to those of AAL1, the following features:

- Efficient bandwidth usage through VBR traffic.
- ATM bandwidth reduction support for voice compression, silence detection/suppression, idle voice channel deletion.
- Multiple voice channels with varying bandwidth on a single ATM connection.

Thus AAL2 can enable voice applications by using higher-level layer requirements such as voice compression, silence detection/suppression, and idle channel removal.

AAL2 is divided into two sub-layers: the Common Part Sub-layer (CPS) and the Service Specific Convergence Sub-layer (SSCS).

a) AAL2 Common Part Sub-layer (CPS)

CPS provides the basic structure for identifying the users of the AAL, assembling/disassembling the variable payload associated with each individual user, error correction, and the relationship with the SSCS. Each AAL2 user can select a given AAL-SAP associated with the QoS required to transport that individual higher-layer application. AAL2 makes use of the service provided by the underlying ATM layer. Multiple AAL connections can be associated with a single ATM layer connection, allowing multiplexing at the AAL layer.

b) AAL2 Service-Specific Convergence Sub-Layer (SSCS)

According to the recommendations of the ITU-T I.363.2, the SSCS is defined as the link between the AAL2 CPS and the higher-layer applications of the individual AAL2 users. Several SSCS definitions that take advantage of the AAL2 structure for various higher-layer applications are planned. A null SSCS, already understood and used in conjunction with the AAL2 CPS, satisfies most mobile voice applications. To satisfy higher-layer requirements associated with data and AAL2 configuration messages—called AAL2 Negotiation Procedures, or ANP—an SSCS for segmentation/reassembly (temporarily called I.SEG) is in development within the ITU-T Study Group 13.

Let's discuss the ATM Trunking standard in brief—an important concept to understand in order to have a successful Voice over ATM network.

a) Switched Trunking

Switched trunking involves analysis of the signaling that accompanies an incoming narrowband call and routing of its bearer information to an AAL2 channel within a VCC between IWFs. Once the narrowband call has ended, subsequent calls occupying the same narrowband channel (TDM timeslot) may be switched to different AAL2 channels and VCCs. In other words, there is no permanent relationship between a narrowband channel and an AAL2 channel.

b) Non-Switched Trunking

In non-switched trunking, the information stream of a narrowband channel is always carried on the same AAL2 channel within the same VCC. In other words, there is a permanent correspondence between a narrowband channel and the AAL2 channel and VCC designated for its support. Non-switched trunking involves no termination of signaling and no routing of narrowband calls in the IWFs.

The interworked voice and data network of the future promises the best of all network worlds: the installed base of Frame Relay, the speed and quality of ATM, and the ubiquity of IP. Currently, fragmentation techniques in Frame Relay, IP, and ATM are quite similar, but prioritization techniques, signaling protocols, and voice compression algorithms are not compatible. Progress is being made toward standardization within each protocol and interworking between them, but considerable work still remains.

Because a comprehensive standard has not been adopted for any one technology, it is unrealistic to expect the emergence of interoperability standards between technologies in the near future. Interworking solutions will therefore have to be proprietary. It is essential that the interoperability be transparent to the users, who want to communicate throughout the network efficiently and without concern for the technological issues involved. Due to the lack of interoperability standards for voice communications over Frame Relay, IP, and ATM, the vendors must develop proprietary interworking solutions.

There are many situations in which interworking between technologies is required within a corporate network. For example, corporations that are running data and voice over Frame Relay might require VoIP to extend the network to remote locations that don't have a Frame Relay infrastructure without deploying additional equipment. This may also be required for telecommuters working from home, salespeople working from hotel rooms, and resellers that want to access information.

Voice over Frame Relay

In the last few years, data networks have been growing at a much faster rate than voice networks, mainly due to the growth of the Internet. Soon the amount of data traffic will exceed that of voice traffic. As a result of this trend, more and more voice is being sent over data networks (Voice over Frame Relay, Voice over IP, and Voice over ATM) rather than data being sent over voice networks (via V.34 and V.90 modems).

When Frame Relay was introduced in the early 1990s, the data technology was not originally designed to carry voice. Despite valid reservations about the reliability of voice over frames, the promise of "free voice" eventually proved too alluring. Soon users were experimenting with transporting voice over their Frame Relay devices while equipment vendors worked overtime to make the promise of quality voice over Frame Relay (VoFR) a reality.

As the public Internet exploded in the mid-1990s and users began implementing IP-based networks, the call for voice over IP (VoIP) grew louder. Here, too, equipment manufacturers are developing products to enable inexpensive, universal voice over data networks.

Carriers, however, were caught in a dilemma. Could they afford to cannibalize their highly profitable public switched telephone network? Could they not afford to capitalize on the demand for digital voice? The drama is just unfolding.

Although significant progress has been made in engineering packet networks (Frame Relay, IP, and ATM) to carry voice as well as data, today's market is demanding a true convergence of these technologies into a single and ubiquitous communications service without being limited by the underlying technology. The next challenge, then, is to develop interconnecting and interworking standards in order to deliver voice services ubiquitously over Frame Relay, IP, and ATM.

Voice over Frame Relay service is enabled by the use of Frame Relay communications devices, such as routers or FRADs, configured with voice modules. These devices are sometimes referred to as Voice FRADs, or VFRADs.

Although VoFR implementations are efficient and cost effective for intra-enterprise or intra-corporate communication, there are considerations, such as quality, to be made regarding the use of packetized voice when dealing with non-Frame Relay WAN, or off-net, users. Figure 2.4 illustrates how a typical Voice over Frame Relay implementation might look, supporting simultaneously both on-net (VoFr) and off-net (POTS) implementations.

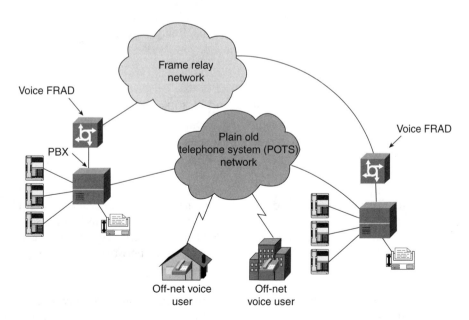

Figure 2.4

Example of Voice over Frame Relay.

Equipping your Company's Network for VoFR

You can deliver voice over any carrier's frame relay services, assuming that you have ensured that the carrier's frame relay network can guarantee a delay and jitter that adequately meets your company's needs. Most of the major carriers offer frame relay services, including MCI WorldCom and AT&T Solutions. Few carriers offer managed voice over frame relay services. However, AT&T Solutions announced a managed voice over frame relay service called *Managed Multiservice Networking* (*MMN*). Unless you use a managed voice over frame relay service, such as that offered by AT&T Solutions, you will have to purchase your own VoFR equipment. Most of the major router vendors, including 3Com Corp., Avaya Inc., Lucent Technologies, Cisco Systems Inc., and Nortel Networks Corp., offer routers, switches, or voice-enabled frame relay access devices (VFRADs).

Voice switching is the technology that enables us to dial a telephone number and reach the desired destination. It implements the signaling required to establish, maintain, and tear down a voice call. Voice switching methods affect the speed, efficiency, and quality of voice calls traveling through the network. Therefore, before purchasing Voice Frame Relay access devices (VFRADs), users must consider the voice switching capabilities of each product, including ease of configuration, flexibility in dialing plans, quality of compressed voice, integrated fax support, and support for various physical interfaces and signaling methods. Before explaining the importance of Voice over Frame Relay, it is necessary to describe the methods implemented to enable the Frame Relay network to carry voice as well as data. In particular, some of the common voice-enabling techniques incorporated into Voice Frame Relay access devices (VFRADs) are described in the following sections.

The Nature of the Data Network and its Implications for Voice

Frame Relay is known as packet-switching technologies. This is in contrast to the public telephone network, which is a circuit-switching technology, designed to carry voice transmissions. Frame Relay inserts data into variable-sized frames or packets.

The packet-switching and cell-switching networks perform statistical multiplexing. That is, they dynamically allocate bandwidth to various links based on their transmission activity. Because bandwidth is not reserved for any specific path, the available bandwidth is allotted according to network needs at any particular time.

Compare this to the traditional voice (or circuit-switching) network, in which a path is dedicated to the transmission for the duration of the call, which is sent in a continuous bit stream. The line is monopolized by a call until it is terminated, even when the caller is put on hold and during periods of silence. Although this guarantees reliable and immediate transmission of voice, it results in very inefficient use of bandwidth. A line that is dedicated to the telephone cannot be utilized by other data even when there are no voice transmissions.

Originally designed to handle bursty data traffic, packet-switching networks (except for ATM) are inherently less efficient than the circuit-switching network in dealing with voice. To achieve good voice quality, the delay of voice packets across the network must be minimal and fixed. Due to the shared nature of the packet/cell-switching network, it might take time for transmissions to travel across the network. A transmission can be delayed because of network congestion. For example, it might "get stuck" behind a long data transmission that delays other packets. Network congestion can also result in dropped packets, which also detrimentally affects the integrity of voice transmissions.

Voice-Enabling the Data Network

Unlike most data applications, voice is very sensitive to delay. Good voice quality provides a faithful recreation of the conversation, with the inflection, pauses, and intonation used by the speakers. Long and variable delays between packets result in unnatural speech and interfere with the conversation. Dropped packets result in clipped speech and poor voice quality. Fax transmissions are even more sensitive to the quality of the transmission and are less tolerant of dropped packets than voice.

One way to deal with the problem of delay and congestion is to add bandwidth to the network at critical junctures. Although this is feasible in the backbone, it is a costly and ineffective solution in the access arena, defeating the "bandwidth sharing" benefits of packet networks. The best solution is to implement mechanisms at the customer premises, access node, and backbone that manage congestion and delay—without increasing bandwidth—such as setting priorities for different types of traffic. Therefore, smart access equipment was developed that could implement procedures to reduce network congestion and the delay of voice packets without adding bandwidth.

Voice over Frame Relay (VoFR) Technology

The significance of Voice over Frame Relay is very important because of its wide existence in the data network infrastructure. To have a flawless converged network with Frame Relay network, we have to ensure several requirements of how Voice over Frame Relay can be deployed. Of the three popular packet/cell technologies (Frame Relay, IP, and ATM), Frame Relay is the most widely deployed. It is commonly used in corporate data networks due to its flexible bandwidth, widespread accessibility, and support of a diverse traffic mix and technological maturity.

Frame Relay service is based on Permanent Virtual Circuits (PVCs). This service is appropriate for closed user groups and is also recommended for star topologies and predictable performance needs. VoFR is a logical progression for corporations already running data over Frame Relay.

Voice Frame Relay access devices (VFRADs) integrate voice into the data network by connecting the router, SNA controller, and the PBX at each site in the corporate network to the Frame Relay network.

Many VFRADs employ sophisticated techniques to overcome the limitations of transporting voice over the Frame Relay network without the need to add costly bandwidth.

Voice over Frame Relay Prioritization

As we know, Voice is a real-time application. Prioritization over any other packet is mandatory to have a successful data/voice converged network. The VFRADs' prioritization schemes "tag" different applications according to their sensitivity to delay, assigning higher priority to voice and other time-sensitive data such as SNA. The VFRADs let the higher-priority voice packets go first, keeping the data packets waiting. This has no negative effect on data traffic, as voice transmissions are relatively short and, being compressed, require very little bandwidth. They can therefore slip into the data network alongside the heavy graphics, payroll information, e-mail, and other data traffic without perceptibly encumbering overall network performance.

Frame Relay service providers have also begun to offer different Quality of Service (QoS). Users can purchase the highest quality of service, Real-Time Variable Frame Rate, for voice and SNA traffic. The second quality Frame Relay service, Non-Real Time Variable Frame, is typically purchased for LAN-to-

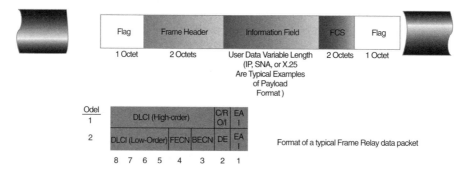

Figure 2.5

A typical Frame Relay Data Packet.

LAN and business-class Internet and intranet traffic. The lowest quality of service, Available/Unspecified Frame Rate, is used for e-mail, file transfer, and residential Internet traffic. In addition, the VFRAD can be configured to assign less sensitive traffic with a Discard Eligibility (DE) bit. These frames will be dropped first in case of network congestion.

Fragmentation

VFRADs incorporate fragmentation schemes to improve performance. Data packets are divided into small fragments, allowing higher-priority voice packets to receive the right-of-way without waiting for the end of long data transmissions. The remaining data packets in the data stream are momentarily halted until the voice transmission gets through.

The downside of fragmentation is that it increases the number of data frames, thereby increasing the number of flags and headers. This also increases overhead as well as reduces bandwidth efficiency. RAD's FR+ application provides an enhanced fragmentation mechanism that fragments data frames only in cases where voice packets arrive at the switch during a data transmission. Otherwise, the long data frames are sent intact.

Controlling Variable Delay on Voice over Frame Relay

Variation in the arrival times between voice or data packets, also called jitter, causes unnatural-sounding voice instead of a smooth voice stream. If a packet does not arrive in time to fit into the voice stream, the previous packet is replayed. This can seriously detract from voice quality. To avoid the effect of jitter, VFRADs detain each packet in a jitter buffer, giving subsequent packets time to arrive and still fit into a natural voice flow. Because the jitter buffer adds to the overall delay of voice transmissions, the optimal jitter buffer should fit the network's differential delay. Also some advance access devices in the market employ adaptive jitter buffering, which continuously monitors the network delay and adjusts the queuing period accordingly.

Voice Compression

Voice compression allows the packet switching network to most effectively carry a combination of voice and data sessions without compromising voice quality. Because Frame Relay access is usually at data rates of 56/64 Kbps, low bit-rate voice compression algorithms such as ITU G.723.1 and G.729A permit the greatest number of simultaneous multiple calls while maintaining high-quality voice. Many vendors have implemented voice compression algorithms in their Frame Relay access devices, and can offer greater bandwidth savings, reduced network congestion, and high-quality voice transmissions.

Silence Suppression on Voice over Frame Relay

In a telephone conversation, only about 50 percent of the full duplex connection is used at any given time. This is because, generally, only one person talks while the other person listens. In addition, voice packets are not sent during interword pauses and natural pauses in the conversation, reducing the required bandwidth by another 10 percent. Silence suppression frees this 60 percent of bandwidth on the full duplex link for other voice or data transmissions.

Echo Cancellation on Voice over Frame Relay

Echo cancellation improves the quality of voice transmissions. It eliminates the echo that results from the reflection of the telephony signal back to the caller, which can occur in a four-wire to two-wire hybrid connection between the VFRAD and the telephones or PBX. The longer it takes the signals to return to the caller, the more perceptible the echo.

Voice Switching Requirements

Various vendors implement voice switching in their VFRADs in different ways. Users should consider the requirements listed following when examining the voice switching capabilities of vendor equipment:

- Flexible numbering plan
- Transparent integration of voice and data
- Easy setup
- Minimal number of virtual links (DLCIs) between sites
- Voice, signaling, and data over a single DLCI
- Efficient switching of compressed voice
- Support and conversion of signaling methods
- Transparent physical interfaces
- Integrated fax support

In addition, the VFRAD should offer the following advanced PBX features:

- Programmable out-of-service indication
- Digit manipulation and DTMF generation (storing and forwarding)
- Hunt Groups

VoFR devices compliant with implementation agreement are not required to negotiate operational parameters. Negotiation procedures are for further study. Therefore, at the time of provisioning, the network manager must configure end-to-end configuration parameters (e.g., Vocoder.). End-point devices providing the VoFR service are configured with compatible sub-channel assignments, signaling, compression algorithms, and other options. The relationship of the Voice over Frame Relay service, VoFR Service user, and the frame relay service is shown in Figure 2.6.

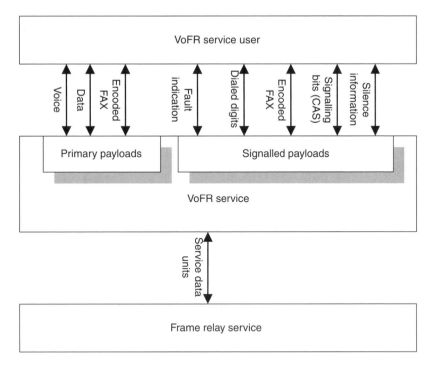

Figure 2.6

Layered architecture of Voice over Frame Relay.

FRF.11 enables vendors to develop standards-based equipment and services that interoperate. It also enables network managers seeking to reduce communications costs and maximize their frame relay network to consider VoFR as an option to standard voice services. In some cases, users may find they have excess bandwidth in their frame relay network that could efficiently support voice traffic. Other telecommunications managers may find that the incremental cost of additional frame relay bandwidth for voice traffic may be more cost-effective than standard voice services offered by local or long-distance carriers.

VoFR can provide end users with a cost-effective option for voice traffic transport needs between company locations. For instance, the network manager may integrate some voice channels and serial data over a frame relay connection between a branch office and corporate headquarters. By combining the voice and data traffic on a frame relay connection already in place, the user has the potential to obtain cost-effective intracompany calling and efficient use of the network bandwidth.

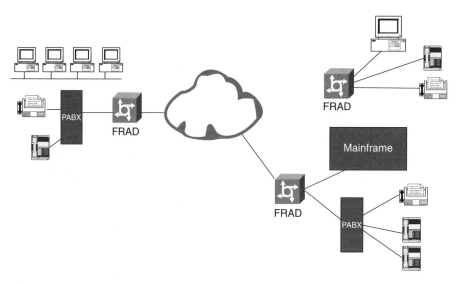

Figure 2.7

Integrated Voice and Data over Frame Relay network.

There are potential trade-offs when implementing VoFR. These include

- loss of the quality commonly associated with toll traffic due to VoFR's use of voice compression
- loss of management and administrative benefits associated with carrier voice services (i.e., the loss of consolidated voice billing and invoice itemization, end user charge back capabilities, and other advanced features such as ID and accounting codes)
- lack of equipment interoperability between the customer premise and the equipment vendor
- lack of standards defining the acceptable levels of quality for voice transport over a carriers' frame relay network.

These trade-offs do not necessarily negate the value and promise of VoFR. Significant advances in digital signal processors and compression algorithms often provide voice at a level approaching toll quality, for a fraction of the cost of public service. VoFR vendors continue to add advanced capabilities in management and administration capabilities. In addition, future industry work will also seek to develop standards which define acceptable levels of quality and performance metrics for voice transport through carriers' frame relay networks.

Voice over Internet Protocol (VoIP)

One of the greatest challenges for Voice over IP (VoIP) is to develop networks that are not only scaleable but also seamless to the subscriber and to the service provider. If the service is difficult to access by the subscriber due to complex dialing plans and special PIN numbers, or requires significant time to complete a call, or has constant call drops, then the IP gateway will only be used by a limited client base. Voice over IP (VoIP) is not a new network, but a new application on IP networks. Traditionally, the voice is transported on a network that uses circuit switching technology, where data networks are built packet-switched technology. There are various reasons this transition is taking place, many of which have to do with economies of scale. Traditionally a Telephony network has been architected around circuit switch technology, requiring specific equipment and management techniques. Networks have emerged from being a difficult to implement, side thought for many companies to a critical part of their business strategy and an integral part of their economic growth.

The overall advantage of VoIP comes from treating voice as another form of data. While claims that the PSTN is dying are premature and unfounded, the advantages presented by IP telephony are clearly visible today.

More Bandwidth

One advantage of VoIP is that it dramatically improves the efficiency of bandwidth for real-time voice transmission, in many cases by a factor of 6 or more. This increase in efficiency is a real long-term driver for the evolution from circuit-switched to packet-switched technology.

New Services

Another advantage IP telephony has over the PSTN is that it enables the creation of a new class of service that combines the best characteristics of real-time voice communications and data processing, such as web-enabled call centers, collaborative whiteboarding, multimedia, telecommuting, and distance learning. This combination of human interaction and the power and efficiency of computers is opening up an entirely new world of communications.

Progressive Deployment

The final advantage of VoIP is that it is *additive* to today's communications networks. IP telephony can be used in conjunction with existing PSTN switches, leased and dial-up lines, PBXs and other customer premise equipment (CPE), enterprise LANs, and Internet connections. IP telephony applications can be implemented through dedicated gateways that in turn can be based on open standards platforms for reliability and scalability.

Market Size

There is a wide range of numbers describing the current size of the IP telephony market and the growth of the market over the next three to five years. While the specific projections vary, even the most conservative analysts are predicting phenomenal growth. The numbers are summarized in Figure 2.8.

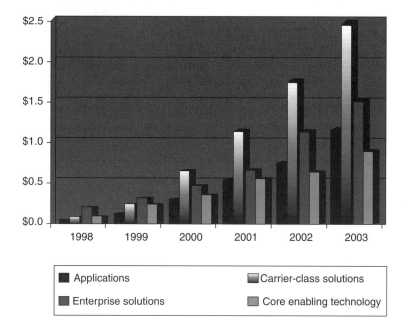

Figure 2.8

IP telephony growth.

VoIP is very complex because it involves components of both the data and the voice world. VoIP is a revolutionary solution as, historically, data and voice worlds have used two different networks, two different support organizations, and two different philosophies. The voice network has always been separate from the data network because the characteristics of voice applications are very different from the characteristics of data applications. The traditional voice network is circuit switched. Interactive voice traffic is sensitive to delay and jitter but can tolerate some packet loss. The voice philosophy was to ensure the "five nines" of reliability—99.999 percent—because the lack of communication might be life threatening (i.e., the inability of placing a "911" call for help). Voice calls have always been networked through dedicated lines to maintain no delay and reliability. The data network, on the other hand, is packet switched. Data is less sensitive to delay and jitter, but cannot tolerate loss. The data philosophy has been concerned with providing reliable data transmission over unreliable media, regardless of delay. Bandwidth in the data world is largely shared, so congestion and delay are often present for multimedia applications such as voice.

Voice and Data network processes work very differently. Data can tolerate delay or latency at a certain limitation, but voice application is very sensitive to delay, latency jittering, and packet loss, and requires much more attention than a data packet. Figure 2.9 shows a typical VoIP converged network with different type media and signaling gateway.

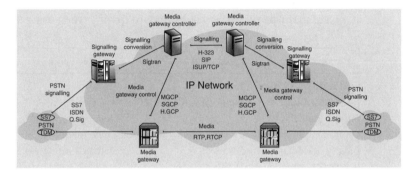

Figure 2.9

Converged architecture.

Media Gateway Controller	Coordinates setup, handling, and termination of media flows at the media gateway.
Signaling Gateway	SS7-IP interface, coordinates SS7 view of IP elements and IP view of SS7 elements.
Media Gateway	Terminates PSTN lines and packetizes media streams for IP transport.

Table 2.5

Implementing VoIP requires attention to many factors, including

Delay

Packet loss

Available Bandwidth

Manageability

Jitter

Packet Prioritization

Latency

Quality of Service

Delay

In VoIP excessive end-to-end delay makes conversation inconvenient and unnatural. Each component in the transmission path—sender, network, and receiver—adds delay. ITU-TG.114 (One-Way Transmission Time) recommends 150 mSec as the maximum desired one-way latency to achieve high-quality voice. Packet delay is the length of time it takes a packet to traverse the network. Users will experience difficulties in carrying on a normal conversation when the one-way network delay exceeds 50 milliseconds (ms). Packet delay in excess of 50 ms can have a noticeable effect. However, some applications or users may elect to tolerate it, just as many people accept substandard quality when using cell phones.

Parameter	Fixed delay	Variable delay
CODEC (G.729)	25 mSec	
Packetization	Included in CODEC	
Queuing delay		Depends on uplink. In the order of a few mSec.
Network delay	50 mSec	Depends on network load.
Jitter buffer	50 MSec	
Total	125 mSec	

Table 2.6

Sample delay budget table

Packet Loss

Typically occurs either in bursts or periodically due to a consistently congested network. Periodic loss in excess of 5–10 percent of all voice packets transmitted can degrade voice quality significantly. Occasional bursts of packet loss can also make conversation difficult. In a VoIP network, there are several factors influencing packet loss.

- Packet loss requirements are tighter for tones (other than DTMF) than for voice. The ear is less able to detect packet loss during speech (variable pitch) than it is during a tone (consistent pitch).
- Packet loss requirements are tighter for short, continuous packet loss than for random packet loss over time. Losing ten contiguous packets is worse than losing ten packets evenly spaced over an hour timespan.
- Packet loss may be more noticeable for larger voice payloads than for smaller ones because more voice is lost in a larger payload.

Available Bandwidth

IP network transmission adds considerable overhead to any given data stream. Each protocol layer adds header and/or trailer information to be transmitted. Depending on their design, IP networks may show hard bandwidth limits or soft limits in which delays and/or losses increase as traffic load increases and congestion builds. For audio traffic, three main approaches are available to minimize bandwidth usage, as follows:

1. choose a low-bit-rate audio codec;
2. combine multiple audio frames into one packet;
3. suppress transmission of silence.

The first method can reduce the audio data rate from 64 Kbps (PCM) to 8 Kbps (G.729), or even as low as 5.6 Kbps (G.723.1). (Note that addition of RTP/IP overhead makes the net traffic differences smaller than the data rate differences, though the use of header compression can mitigate this effect.) Two sacrifices must be made to obtain this reduction: Processing must be expended to perform the encoding and decoding; and, the audio quality is degraded to an extent dependent on the codec used. The second method reduces packet overhead, does not affect audio reproduction, but does increase the latency by the time represented by the number of additional frames in each packet. The third method can reduce overall audio traffic by roughly half, but the success of this effort depends on the statistical averaging effects of many calls. Another approach to bandwidth management is through prioritization: Give audio preferred access to the limited facilities and let other types of data streams contend for what's left. This approach may be invoked by choice of an 802.1p priority (at layer 2), or a DiffServ service tag (at layer 3).

Despite the high speeds available in today's LANs, they are not immune to QoS issues. Shared media, for example, can cause problems if collisions occur (because at least one user may need to be delayed). Prioritization of traffic on the LAN allows QoS requirements to be signaled to LAN switches and routers. The IEEE ratified a standard for QoS prioritization in IEEE802 LANs in 1999. The IEEE802.1p specification defines three bits within the IEEE802.1Q field in the MAC header (which is a part of OSI Layer 2). The IEEE802.1Q field was initially designed to support VLAN interoperability but has been extended to support traffic priorities. This information can be used as a "signal" to any device that can decode the bits. Figure 2.10 illustrates the approach.

Three bits allows up to eight settings that can be used for classes of traffic and priorities. Typically, a NIC card in a LAN-attached system sets the bits according to its needs, and this information can be used by Layer 2 switches to direct the forwarding processes. If multiple distinct LANs are interconnected via routers (i.e., Layer 3 switches), then the Layer 2 bits must be used to drive Layer 3 QoS mechanisms. Although the IEEE802.1p/Q mechanism cannot operate on an end-to-end basis in an internetwork, it does provide a relatively simple method of defining and signaling an end system's requirements within a LAN environment. Multimedia applications and convergence changed the situation—usage and users must now be distinguishable and network processing needs to be tailored to each class of traffic. For example, a real-time telephone call should

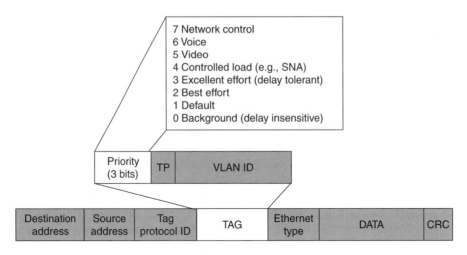

Figure 2.10

IEEE802.1p/Q prioritization bits.

not be handled the same way as a large file transfer if they share a link. Differentiating among QoS requirements can be done by classifying each packet and using this information in switching decisions. The IETF (in a working group called DiffServ) is completing a series of standards that re-define the ToS byte and re-name it as the Differentiated Services Control Byte (DSCP). RFCs 2474 and 2475 provide the details. Figure 2.11 illustrates the DSCP within the context of the IP protocol header.

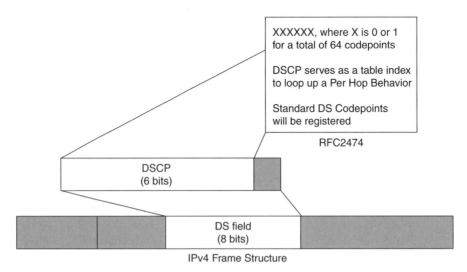

Figure 2.11

Differentiated services bits.

DiffServ [RFC2474] makes use of the existing Type-of-Service (TOS) octet in the existing IP Version 4 header [RFC791]. As such, it may be set by information senders and used by IP (Layer 3) routers within the enterprise network. Effectively, the TOS octet is really just 6 bits wide because neither DiffServ nor IP TOS use the least-significant 2 bits; they are reserved for future use and fixed at 00. The original TOS definition was: bits 0–2 Precedence (7 = highest, through 0 = lowest) bit 3 Delay (0 = normal delay, 1 = low delay) bit 4 Throughput (0 = low, 1 = high) bit 5 Reliability (0 = normal, 1 = high) bits 6–7 reserved for future use (set to 00) DiffServ combines bits 0–5 into a single selector for "Per-Hop Behavior" (PHB).

Compatibility with older systems using precedence is preserved by fixing the 8 PHB values of the form xxx000 (binary) to be equivalent to the behaviors provided by networks that provided precedence routing; in particular, the value 000000 must always represent the default "best effort" service provided by each IP network. Generally, the 6 PHB bits represent 64 "code points" or possible per-hop behaviors. These 64 are divided into 3 sets: xxxxx0 for (future) standardization; xxxx01 and xxxx11 for experimental or local use. Not all values may be supported by all the networks, and it may be necessary to map the PHB values from one network to another to obtain a relatively consistent behavior. Until PHB values are standardized across networks, and all networks implement DiffServ, each endpoint on a network must be configured to use the value(s) appropriate for that network's traffic, and let the inter-network gateways translate, if required. In the absence of any prior knowledge, a participant in an IP conversation could start with the default PHB codepoint (000000), then move to another when it discovered a different codepoint being received with the same type of traffic from the other party to the conversation. Once the codepoint is changed from 000000, however, this procedure should be abandoned to avoid constant re-mapping if the intervening networks do not have a strictly reversible mapping.

Manageability

On a converged network, to have a seamless performance it is crucial to manage your network carefully and efficiently manage your limited bandwidth, particularly for real-time applications such as voice packets over the network. One of the options to manage your network in an efficient manner is a policy-based management tool with a real-time voice/data monitoring tool. Usually an excellent class policy based with real-time voice/data and converged network management system supports configuring the network in accordance with business needs and requirements. Most network vendors offer or support PC/Windows applications that provide a means to express business policies in terms of rules that govern the behavior of the network. These rules may, in turn, be expressed in terms of users, applications, the time of day, and monitor real-time applications. Policy-based network rules configure (and reconfigure from time to time, as necessary) the network so that its behavior carries out the business needs. It implements the rules by configuring individual network devices to behave in accordance with the network policies, typically utilizes a variety of protocols, such as telnet/Command Line Interface, COPS, and SNMP, to communicate with network devices to achieve the desired configuration. The way policy-based network management works, the rules are generally stored in a directory with a schema designed to support rule-based policy management. Each IP switch and router in the converged enterprise network should be able to be configured to provide a variety of services. These services provide low latency and constant delay as is required for good voice quality. This service is generally configured by assigning a particular queue(s) in the device that is to be used for this specific traffic. This traffic may be designated as that originating from one or more specific physical ports on the device, that originated from a UDP port within a specified range, or as that tagged with specific 802.1p/Q and/or DiffServ code point values. The assigned queue(s) is(are) then given priority over other queues that are used to forward latency-insensitive traffic. Though all of these mechanisms may be utilized under control of the policy-based management system, only UDP port ranges and packet tagging via 802.1p/Q and DiffServ are flexible enough to be used entirely under software control interests and are the services that provide low latency and constant delay as is required for good voice quality.

Jitter

Perhaps the one aspect of IP networks that makes them fundamentally different from circuit-switched networks is variance in propagation delay, or jitter. Audio encoding/decoding is basically a synchronous process: The analog signal is sampled at prescribed intervals, the samples (or their encoding) are transmitted, and the received samples are clocked out to reproduce the analog signal. IP networks are basically asynchronous transport devices. Clocks at the sender and the receiver are not synchronized, hence the receiver may play out audio slightly faster than it was generated (thereby sometimes running out of audio to play) or slightly slower (thereby falling further and further behind the source).

To make matters worse, individual packets of audio may experience more or less delay as they travel through the network. This causes variations in the inter-arrival time. Some packets may experience "infinite delay" (be lost) or such long delay that they become useless or stale. The main technique for dealing with variable propagation delay is the jitter buffer: Incoming audio is passed through the jitter buffer and some amount of audio information (PCM samples, say) is buffered up before playing the samples out. If an incoming packet is delayed somewhat, then the information already in the buffer can be played out until either the buffer empties or the information arrives. Because the packets are pre-

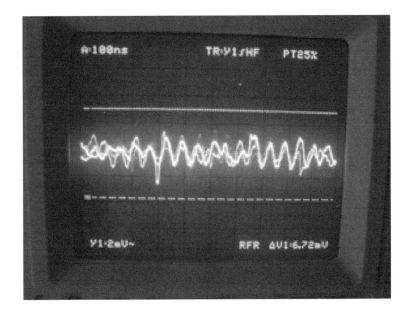

Figure 2.12

Sample jitter testing: What we see on the scope is not the signal that carries the jitter, but the jitter inside the signal.

sumably coming at a fixed rate, when one or more packets arrive late, some subsequent packets should arrive "early" (or on time) to refill the buffer. If the buffer runs dry, some type of audio fill-in must be supplied. As network jitter gets worse, the size of the jitter buffer must be increased to avoid too many under-runs. Unfortunately, the use of a jitter buffer introduces a delay proportional to the buffer's target size. Some products (e.g., Microsoft NetMeeting) create an extremely large jitter buffer on the order of 200–300 ms in order to handle jitter from virtually any type of connection, which introduces a "fixed" delay degradation of the conversation.

Voice Packet Prioritization

Prioritization of network traffic is simple in concept: Give important network traffic precedence over unimportant network traffic. One prioritization scheme assigns priority based on the UDP (User Datagram Protocol) port numbers that the voice packets use. This scheme allows you to use network equipment that can mark the packets on these ports with a priority. UDP is used to transport voice through the LAN because, unlike TCP, it is not connection-based. Because of the human ear's sensitivity to delay, it is better to drop packets rather than retransmit voice in a real-time environment. Prioritization is also called CoS (class of service) because traffic is classed into categories such as high, medium, and low (gold, silver, and bronze), and the lower the priority, the more "drop eligible" is a packet. E-mail and Web traffic is often placed in the lowest categories. When the network gets busy, packets from the lowest categories are dropped first.

Prioritization/CoS should not be confused with QoS. It is a subset of QoS. A package-delivery service provides an analogy. You can request priority delivery for a package. The delivery service has different levels of priority (next day, two-day, and so on). However, prioritization does not guarantee the package will get there on time. It may only mean that the delivery service handles that package before handling others. To provide guaranteed delivery, various procedures, schedules, and delivery mechanisms must be in place. For example, Federal Express has its own fleet of planes and trucks, as well as a computerized package tracking system.

Prioritization has been used in multi-protocol routers to give some protocols higher priority than other protocols. For example, SNA (Systems Network Architecture) traffic will time-out if it is not delivered promptly, causing retransmissions that degrade network performance. Such protocols should be given high priority. A number of other prioritization/CoS schemes are outlined here.

- MAC layer prioritization In a shared LAN environment such as Ethernet, multiple stations may contend for access to the network. Access is based on first-come, first-serve. Two stations may attempt simultaneous access, causing both stations to back off and wait before making another attempt. This is minimized for switched Ethernet where only one station is connected to a switch port. A number of vendor-specific Ethernet priority schemes have been developed. Token ring networks have a priority mechanism in which a reservation bit is set in tokens to indicate priority.
- VLAN tagging and 802.1p The IEEE 802.1Q frame-tagging scheme defines a method for inserting a tag into an IEEE MAC-layer frame that defines membership in a virtual LAN. Three bits within the tag define eight different priority levels. The bit settings serve as a label that provides a signal to network devices as to the class of service that the frame should receive.
- Network layer prioritization The IP packet header has a field called ToS (Type of Service). This field has recently been redefined to work with the IETF's Differentiated Services (Diff-Serv) strategy. Diff-Serv classifies and marks packets so that they receive a specific per-hop forwarding at network devices along a route. The ToS bit is set once, based on policy information, and then read by network devices. Because IP is an internetworking protocol, Diff-Serv works across networks, including carrier and service provider networks that support the service. Therefore, Diff-Serv will support CoS on the Internet, extranets, and intranets.

Priority settings may be made in several places. The most logical place is the application running in the end user's system. But applications may not support the various schemes that are available, meaning that edge switches may need to infer priority levels for frames or packets by examining the contents of the packets. This is now easily done with so-called "multilayer switches" based on policies that are defined in policy-based management systems.

Latency

The average time delay experienced between an audio source (speaker) and destination (listener) does not affect the quality of speech directly, but it affects the speaker's perceptions of echo from the listener's end, and it affects the quality of a conversation (the ability of the listener to "turn the speech path around" and respond). The most practical means of treating the echo problem is to cancel it at the endpoint closest to the cause of the echo. For an IP endpoint (e.g., telephone set), this means within the set itself. Endpoints with speakers and microphones can be troubled by acoustic echo. When an IP connection is inter-worked to the circuit-switched network, it is necessary to cancel any hybrid-induced echo that might be generated from a circuit-switched endpoint. This cancellation should take place in the interworking device. Telephone users expect good conversational quality on a call. Only in the most extreme cases (Earth to Moon) do they accept long delays without question. Frequent users of calls relayed via satellite have some acceptance of long delays, and are practiced in the required protocols, but most users are unwilling to accept such inconveniences on calls that are essentially "local." Although there is no specific breakdown point, the generally accepted standard is that one-way delays in excess of 150 ms or round-trip delays in excess of 300 ms are intolerable. Perhaps the most obvious contributor to latency is the audio packet propagation delay across the network (although it is often not the major contributor). Propagation delays (including media access delays) can vary from a few milliseconds on a LAN to hundreds of milliseconds and are strongly dependent on network loading and topology. The delays due to loading can be reduced by giving audio packets priority service over other network traffic. Various methods of prioritization are in use by vendors: explicit priority labeling (e.g., 802.1p or DiffServ), or via address recognition (source/destination address preferences or port number preferences). Audio compression and packetization also contribute to latency. Most coders operate on an audio sample interval or frame, typically 10 ms. One or more encoded frames is then assembled into a packet for transmission (multi-frame packets reduce overhead). This represents a delay of one or more frame times. So it is crucial to monitor and control latency on your converged network.

Quality of Service

Virtually all modern networks use a layered protocol structure with each higher layer taking on more responsibility for successful end system interworking. QoS mechanisms can be incorporated at any layer (or, in fact, in all of them).

Table 2.7 illustrates the five layers of the TCP/IP stack and the positioning of the key QoS mechanisms.

Application Layer	User & application access, profile authorization, encryption, QoS-aw intelligent applications design for QoS
Transport Layer	Port-based access control, TCP rate control
Network Layer	Resource reservation, Type of Service bits, path controls, load balancing, address-based access control, protocol-based separation
Data Link Layer	Frame prioritization (802.1p/Q), ATM Cell QoS, VLAN and VPN isolation, logical port access control, path controls
Physical Layer	Bit error correction, link diversity, path controls, physical security, physical port access

Table 2.7

Note that some of these mechanisms are inherent in the protocols (e.g., priority based on physical ports, decisions based on Layer 4 port numbers) rather than being explicitly added on for QoS control.

Most companies have become dependent on their networks. Who, for example, would be able to prosper in today's society without a telephone (even for calling someone in the same office)? Imagine the inconvenience if the telephone system was frequently so busy that you could not even get a dial tone or if a telephone call suddenly failed in the middle of a conversation? The same concerns arise when data networks are business critical, such is true for a bank teller or on-line book ordering. There can be little doubt that the quality of the network is very important, almost as important as quantity in some cases. The lack of adequate QoS can rapidly become a business inhibitor. Why then is there a sudden interest in dynamic QoS management for enterprise networks? Nothing about the traffic or its value has changed, although new forms of digitized traffic (such as video) have become available. What is different is that the underlying enterprise network technology is changing (to the TCP/IP architecture pioneered by the Internet) and the desire is now to "converge" all forms of traffic onto a single TCP/IP-based infrastructure. Convergence is aimed at reducing costs and complexity, increasing operational and managerial efficiency, and opening the door for innovative new multimedia applications. The original TCP and IP protocols were designed for applications such as e-mail distribution, basic file transfer, and remote terminal access, none of which were particularly time or bandwidth sensitive. The primary goal was to establish connectivity across a wide variety of systems and devices using a robust, inexpensive network. The philosophy was to assume that the network would do its best to deliver all messages that were submitted but with no guarantees of delivery or sequencing. This "best efforts" approach is quite satisfactory for basic data communications (and hence the popularity of the Internet) but has proved less suitable for networks that must also accommodate voice and video traffic (i.e., multiservice networking). The introduction of IP-based QoS mechanisms and the extension of network management systems is an essential step in the transition to the new, converged network infrastructure. Multiservice traffic is difficult to handle efficiently by very nature, especially in resource-limited environments.

The goal of the QoS strategy is to provide tools and mechanisms that allow the network administrator the means to optimize the performance of the network. This often involves tradeoffs between competing factors. For instance, basic PCM audio encoding is, by definition, a synchronous process; IP network transport is, by design, an asynchronous process. Traditional circuit switching provides dedicated, clocked resources for lockstep transmission; packet transmission is subject to unpredictable delays and losses. The network administrator's job is to use the QoS features to minimize the effects of IP network "unpredictability."

In a converged network, the voice call/feature server plays a central role in ensuring that the available QoS mechanisms are appropriately utilized so that voice traffic receives appropriate treatment in the network consistent with business policies regarding network utilization. This section describes the QoS mechanisms available in a converged network and how the call server controls their use. IP-based QoS mechanisms can be utilized to provide the best possible end-to-end audio experience when all or part of the audio path is carried over packet-switched networks. "Best," in this context, is defined by the customer as represented by the network administrator, and represents a tradeoff between audio reproduction quality, audio path delay (latency), audio loss, and network resource consumption. The network administrator is able to effect this tradeoff by means of audio codec selection and invocation of network prioritization through the Differentiated Services (DiffServ) capability [RFC1889, RFC1890], and/or the IEEE 802.1p/Q MAC-layer prioritization and segregation. Use of DiffServ and/or 802.1p/Q priorities applies to either or both ends of a voice path: The voice call server configures media processor resources and the terminals involved in the call to use the appropriate values for the (sub)network(s) to which they are attached. Typically, these values will be uniform throughout the network, but allowances are made for different parts of the network to utilize different DiffServ and/or 802.1p/Q values to designate the same service.

In order for voice traffic to receive excellent quality in a converged network, the network infrastructure equipment, i.e., industry standard IP switches and routers, and the voice endpoints, which include the telephones, softphones, and gateways, must be configured in a consistent manner so that voice packets are appropriately tagged and, consequently, properly identified and forwarded expeditiously. QoS architecture employs a policy-based network management system to consistently configure the network and the voice call/feature server with QoS parameter values. The voice call/feature server is, in turn, responsible for conveying these values to the voice endpoints.

CHAPTER 3

THE IP-BASED CONVERGED NETWORK ELEMENTS

One of the most important aspects of the modern, IP-based converged network architecture is that necessary functions can be broken out into logical components. In this way, scalable and interoperable solutions can be constructed to meet the different needs of the different service providers, enterprises, or end users at a reasonable cost, while enabling services to be supported locally as well as across an entire network. Some of the advantages of this approach are

Users can improve the speed of deployment of their solution through off-the-shelf network elements. When these elements are developed to adhere to published and agreed-upon international standards, they can be custom fit into unique, yet interoperable and scalable solutions.

Competition among equipment suppliers is promoted by the demand for open standards. This keeps costs low and drives innovation so that the prohibitive cost barriers to introducing advanced network services are significantly reduced.

Separating the control and media elements enables the rapid development of advanced applications such as advanced calling features, multimedia contact centers, unified messaging, and future developments such as presence and intelligent agents.

This approach enables the evolution of the fully managed, converged IP network that supports all of the separate applications and enables integration with the Internet and the Public Service Telephone Network (PSTN), which further improves the range of available services.

In this chapter we will discuss, in some detail, the basic building blocks of a converged, IP-based voice, video, and data network. In actual implementations some or all of these elements may be combined into a single physical device or server, but for the purposes of this discussion we will treat them all as separate entities. There is really no consensus in the industry on the exact, generic terms for these elements and, to add to the confusion, many product vendors use the generic names as their product name. For our purposes we will use the following basic converged network element names as generic terms:

Managed IP Infrastructure: This is the core infrastructure that transports the data, voice, and video packets. The IP infrastructure also provides all of the mechanisms to assure reliable transport and quality of service (QoS).

Media Gateways: Media Gateways are physical devices that translate between one media type and another. In a discussion of IP-based converged networks, the media gateway is typically a device that converts traditional telephony end-points to IP and vice versa. There are also specialized media gateways that are typically referred to as **Access Gateways**, **Trunking Gateways**, and **Network Access Servers.**

Signaling Gateways: Signaling gateways are specialized media gateways that convert public telephone signaling channels such as Signaling System 7 (SS7) into IP.

Media Controllers: Media Controllers (MC) manage the various media gateways and terminals in a network and provide for end-to-end call control across the network. The MC also controls access to the network for voice services and controls the registration and access of end devices. MC also controls addressing and feature access and provides mechanisms for call tracking and billing. MCs are known by many names in the various industry standards, including **Gatekeepers**, **Call Agents**, **Softswitches**, or **Telephony Servers**.

Multipoint Controller Units: Multipoint Controller Units (MCU) are used in an IP environment to connect multiple parties (more than two) into a single connection. MCUs are typically used to provide voice or video conferencing services.

Adjunct Processors: Adjunct processors are devices that provide peripheral services to the converged network. These processors can be items such as **Network Management Servers**, **Advanced Applications Servers** (e.g., Contact Center Applications or CTI Servers), **Call Detail Recording Servers**, and many others.

Terminals: In this context "terminal" refers to an IP-based endpoint device such as an IP Telephone, Video Unit, or Softphone. IP terminals connect to each other and to the gatekeepers/softswitches across the IP network using the H.323 protocol or Session Initialization Protocol (SIP). IP terminals connect to non-IP stations and networks by means of a Media Gateway.

Figure 3.1 shows a generic IP-based converged network.

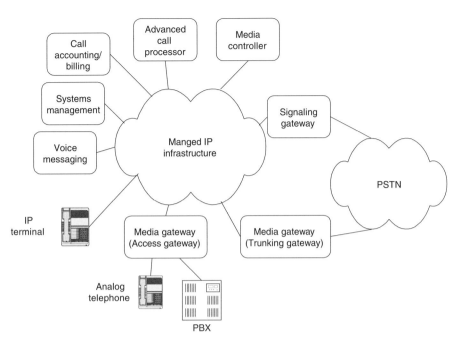

Figure 3.1

Typical IP-based converged network.

Managed IP Infrastructure

The basis of a converged network is the underlying IP Infrastructure. The IP infrastructure is vital to ensuring that the various services of data, voice, and video are handled as efficiently and effectively as possible.

The IP Infrastructure provides all of the physical transport of IP voice, data, and video along with all of the signaling and control information needed to ensure the delivery of those services.

A managed IP Infrastructure also includes the network management tools that are used to assure the function of the infrastructure. Specialized management tools are used to ensure Quality of Service (QoS) across the IP infrastructure, vital to the reliable and efficient handling of multiple media such as voice data and video.

Throughout this book we will be looking into the many issues regarding a multiservice-ready IP network. These issues include providing reliable and scalable transport of information, maintaining QoS across the local area and wide area networks, and managing, maintaining, and controlling the network for optimal performance.

Media Gateways

The Media Gateway (MG) is the basic physical interface for an IP-based converged network. The media gateway translates between networks of differing standards. It provides conversion of streamed media formats such as voice or video, and manages the transfer of information between the different networks.

The typical media gateway in an IP-based system has multiple physical interfaces. These interfaces usually include

- *IP interfaces (usually Ethernet)* to connect to the packet-switched network.
- *Analog Station Interfaces*, often called Foreign Exchange Stations (FXS), to connect directly to traditional analog telephones, voice messaging ports, or analog voice recorders. Often these types of interfaces are referred to as Tip/Ring (T/R) interfaces.
- *Analog Trunk Interfaces*, often called Foreign Exchange Off-nets (FXO), to connect to analog PSTN interfaces such as central office (CO) trunks, direct inward dial (DID) trunks, and other traditional analog network interfaces.
- *Analog Tie Trunk Interfaces*, to connect via analog facilities to other media gateways or to traditional voice equipment such as PBXs or key systems. These interfaces are often referred to as E + M (ear and mouth) tie trunk.
- *Digital Trunk Interfaces*, to connect digital PSTN facilities such as T1 or E1 interfaces to a network service provider or to interface to other media gateways or traditional voice equipment.
- *Wireless Network Interfaces*. These interfaces include 802.11 or WiFi interfaces as well as public wireless interfaces such as GSM.
- Specialized interfaces are also sometimes included, such as interfaces to ATM network services or SONET.

The media gateway is the device that provides not only the physical interfaces to these types of networks and devices, but also translates between the protocols used on those interfaces.

As an example we will look into the functions of a media gateway that connects analog stations to an IP network. In this example the media gateway needs to take in voice signals from the IP network and translate them into analog signals on a traditional analog phone. This means that the media gateway is taking in the IP packets containing the voice information, stripping off the IP protocol information, and re-coding that information back into an analog signal. The media gateway must also perform that conversion in the opposite direction, from Analog to IP. Figure 3.2 shows an illustration of this example.

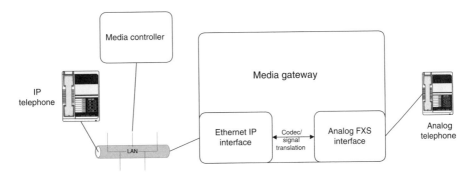

Figure 3.2

Analog FXS media gateway.

The media gateway provides these functions through internal Coder/Decoders (codecs). Common codecs supported by a typical media gateway include G.711 (64 Kbps voice), G.729 (8 Kbps voice), and G.723 (5.3–6.3 Kbps voice).

In addition to translating the voice or video signal, the media gateway must also perform the same translation between the signaling for the various interface types. In the preceding example, this would involve translating IP messages into analog signaling and back again. For instance, the media gateway would translate the IP message for Call Offered into ringing on the analog station.

This means that a media gateway is also responsible for providing signaling mechanisms such as call progress tones, dial tones, DTMF, and others. The media gateway must also have facilities to detect and often collect these tones on the analog interfaces and translate them into the appropriate protocol to be used on the IP network.

Collected analog messages such as DTMF tones can be transported across the network in one of two ways. They can be encapsulated in the IP stream as tones, or more commonly, they can be translated into equivalent signaling messages in the protocol. Once these messages are translated into an IP format, they are then sent to a media gateway controller (MGC). Based on this information and its internal connection rules, the MGC will then inform the media gateway as to how to process the call.

The media gateway must perform similar functions to translate between any of the physical interfaces on that particular gateway. So a media gateway may transform H.323 or SIP messages into T1 signaling and back, or it may translate incoming connections on a digital tie facility from a PBX into IP messages and packetized voice.

Media Gateways are also used within an IP network or between two different zones in an IP network to translate between protocols or codecs. This is often used to connect two different parts of the network that may have different needs. An example of this is a network that uses a G.711 encoding for IP voice on the local area network but needs to translate that into a more compressed coding such as G.729 to transport across a more bandwidth-sensitive WAN link. This translation is known as *transcoding*. In this way the media gateway acts as a shared resource for digital signal processing in the IP network.

Another example where a media gateway may be used between two different regions of a packet network is when two different voice-over-IP protocols are being used such as translating H.323 into SIP. This function of the media gateway allows for the phase migration of the network into new technologies. That is one of the biggest strengths of this decentralized architecture in that it allows newer technologies to interwork with older technologies through the use of standards-based interfaces.

A number of specific media gateway types are defined according to studies performed by ETSI (Tiphon) and the IETF. Some of these types are

Access Gateway
Access Gateways connect user network interfaces (such as ISDN or traditional analog services) to an IP. As such, it will typically terminate TDM call signaling and pass this information to a Media Gateway Controller for call control decisions to be made.

Trunking Gateway
Trunking gateways interface between the PSTN telephone network and the IP-based network. Such gateways typically manage a large number of digital virtual circuits and TDM bearer circuits with signaling carried on a separate path (through a signaling gateway).

Network Access Server
The NAS is a specialized form of the Access Gateway designed to terminate modem calls or HDLC connections and provide data access to the IP network.

Media gateways are typically under the control of a central processor that makes overall routing and call handling decisions. This processor is known as a Media Controller (MC).

Signaling Gateways

The Signaling Gateway (SG) is a specialized gateway that is responsible for termination of Switched Circuit Network (SCN) signaling (typically SS7) and transport of signaling messages to the MGC across the managed IP network. It also allows remote devices on the IP network to exchange messages with the PSTN network for call setup or for querying SCN database servers that support advanced network services such as local number portability, toll free numbers, etc. The SG implements the functions and interfaces defined by the SIGTRAN standard.

As typically defined, the signaling gateway is more of a specialized device for service providers and very large enterprises and is used to interface directly with the public networks Signaling System 7 (SS7). The purpose of the SG has been to maintain the separation of signaling links and voice bearer channels that is a core part of the public switched network architecture.

Lately, this architectural concept is being expanded into the enterprise network space. This expansion is happening by developing specialized gateways, much like signaling gateways that translate enterprise signaling protocols, most typically ISDN-PRI, into IP messages and back again. By separating this function out of the traditional PRI media gateway, the ISDN messages can be transported separately from the voice bearer links.

An example of this is shown in Figure 3.3.

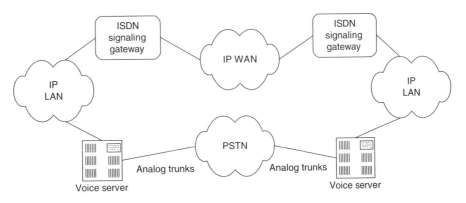

Figure 3.3

Separation of ISDN signaling over WAN.

In this example, the ISDN messages are translated into IP and transported to the far end over an IP WAN. The voice conversations are left in an analog state and transported across public telephone facilities.

ISDN messages between facilities can provide advanced calling features such as passing calling number and names down to the station level. A protocol called Q-Signaling (Q-SIG) also provides for even more advanced features such as centralized attendant consoles and centralized messaging with far-end message waiting indication.

By separating the ISDN and Q-SIG messaging from the physical trunk, the network can take advantage of these advanced features without using expensive digital facilities and without increasing the IP WAN's bandwidth to support the voice traffic. The main challenge in this implementation is re-associating the calls to their ISDN messages at the far end. Many different vendors have proposed schemes to overcome this challenge and make this promise of IP telephony services more viable.

Media Controllers

The Media Controller (MC) is the core call-processing element in an IP-based network. The MC is responsible for all of the call processing functions across the zone of the network that is under its control.

One of the major difficulties in understanding IP-based telephony is that there is no one, overall, standards body controlling standards development in the industry. There are actually many different standards groups involved in this development. Groups like the ITU, IETF, ECMA, IEEE and many other groups are responsible for defining the standards used for IP telephony. A major part of the confusion caused by this diversity of standards bodies is there is no consistent definition of terminologies. Nowhere is this inconsistency more noticeable than in the nomenclature surrounding the Media Controller. Media Gateway Controller is the name used for this device in the Megaco/H. 248 standard. The same functionality is defined in the MGCP protocol as a Call Agent. H.323 refers to the device as a Gatekeeper. Some product developers call this device a Media Server, a Telephony Server, or a Call Manager. Just to add to the confusion many network models break the functionality described here into different parts, referring to the overall call control as an MGC and the registration and physical media control as a Gatekeeper. There are many variations on this theme. For purposes of this discussion we will use the term media controller or MC as a generic term as it includes all of the functions typically described to a media gateway controller and a gatekeeper while leaving the terminology open to discuss terms that apply to many different protocols.

The Media Controller, as we said, is the core management tool available for an IP multimedia network. As the brains of the network, this application performs essential control, administrative, and managerial functions required to maintain the integrity of networks in both enterprise and carrier environments. The MC provides the following functions:

- authentication
- authorization
- accounting
- call control and call routing

- basic telephony services such as directory services and private branch exchange (PBX) functions
- control of bandwidth usage to provide quality of service (QoS) and protect other critical network applications from traffic
- total network usage control
- overall system administration and security policies

The MC is the focal point of the multimedia network. IP telephony standards are typically implemented in production networks through subdivisions known as zones or regions. Zones are the set of endpoints over which one and only one MC has jurisdiction. They include any number of terminals, gateways, and multipoint conferencing units (MCUs); any combination of these entities may register with the gatekeeper functionality of the MC. Regardless of the physical location of the program code, there is usually only one active MC per zone. Zones can be defined according to geographic locations (such as different branch locations) or in accordance with overlap of a physical network connection (such as a subnet on the floor of a building or a range of IP addresses) or by a functional (organizational) paradigm (see Figure 3.4).

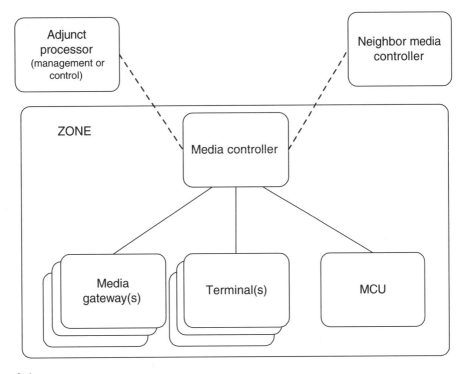

Figure 3.4

Typical zone configuration.

The MC manages all zone activities. Whenever an entity goes live, it will either send out a query to the network, asking which (if any) MCs are present and willing to accept a registration request from this entity or send a registration request to a predefined MC. This endpoint discovery and registration process is a prerequisite for zone management.

The choice of MC is critical to the optimal operation of a total solution. The MC allows developers the ability to scale up the system to a large number of users while taking care of interzone call routing.

Identifying endpoints in a zone is done using IP addresses, alias names (such as H.323 identifiers), e-mail addresses, and universal resource locators [URLs]) or phone numbers. The MC is the focal point for insertion of logic into the network. It can be configured and controlled remotely by third-party applications using hypertext transfer protocol (HTTP) or signaling network management protocol (SNMP). For example, a network planner can configure an MC to allow a specific set of endpoints into the zone and provide users with unique policies and procedures.

Specifications defining the core role of the MGC include the following:

- address translation
- admissions control
- bandwidth control
- network management

Address Translation

The MC provides address translation between alias and transport addresses upon an endpoint's request for service. As users typically do not know the IP addresses of other terminals (or entities) they wish to call, the MC translates an alias address (H.323 identifier, URL, phone number, or e-mail address) to a transport address. There are several mechanisms that can be implemented to update the translation table, one of which includes using the registration, admission, and status (RAS) channel. Other mechanisms can be implemented to support endpoints that do not use the RAS channel for routing communications messages.

Admissions Control

The MC authorizes network access based on network guidelines and other criteria using admissions request message, admissions confirm message, and admissions reject message. As a result of limited share resources, not all users will be able to access the network at the same time. MCs also protect the integrity of the network for all users according to specified policies. MCs will authorize network access along the policy guidelines that a network administrator has selected when setting up a MC and/or zone. Network access and special services (use of a gateway, for example) may be based on call authorization, bandwidth usage, or other criteria.

Bandwidth Control

The MC monitors and controls network bandwidth usage and ensures that audio and/or video traffic does not exceed maximum network load as defined by the network manager. Network managers have the ability to restrict usage for traffic to offer quality of service (QoS) to other more critical applications. The MC may also accept all requests for bandwidth changes where no policies are enforced.

Network Management

The MC uses a routed call-signaling model to route call signaling and control channels to the appropriate entities in the network. Furthermore, the MC can implement logic for granting/denying terminals, gateways, and MCUs access to the associated network assets such as bandwidth, gateways, MCU, directory services, etc. It performs this procedure by monitoring all concurrent calls in a zone and enforcing network management policies for any new calls (sessions) a user may initiate.

In an effort to differentiate their products from their competitors, network designers can customize their MC by implementing the following optional features:

- call control routing
- call authentication
- call authorization/access
- call accounting
- bandwidth management
- call management services
- supplementary services
- directory services

Call Control Routing

There are two models for call routing: direct mode and routed mode. The routed mode is the more widely preferred model. When the MC performs address translation, it provides endpoints with the transport address for the call signaling channel destination. In the direct mode, the MC provides the endpoints with the address of the destination endpoint and directs them to the call-signaling channel so that all messages can be exchanged directly between the two endpoints without MC involvement.

With the routed mode, the MC provides its own address as the destination address so that it receives all call-signaling messages and handles routing the call signals between itself and all endpoints during a session. In this case, the MC keeps a signaling channel open while routing the call for the duration of the call. The routed mode is fundamental for call management, as it performs line-hunting functions, provides separate control over each leg of the call by disconnecting and reconnecting each leg separately, and provides supplementary and proprietary services. This MC-routed model is the preferred method for ensuring the efficient delivery of supplementary services as well as a more robust management of the network (see Figure 3.5).

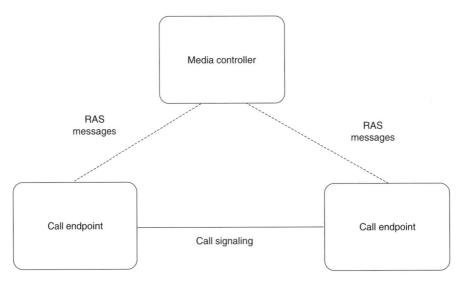

Figure 3.5

Call signaling in the routed mode.

Call Authentication

The MC has the capability to perform call-authentication functions by identifying the user or providing a given token or certificate.

Call Authorization/Access

The MC authorizes a call based on the user's access rights. It may reject calls from a terminal as a result of authorization failure. The reasons for rejection may include but are not limited to restricted access to or from particular terminals or MCs and restricted access during certain periods of time. Corporate management information system (MIS) and information technology (IT) management have the flexibility to determine the criteria as to whether or not authorization passes or fails, based on security reasons such as restricted access to services or out-of-zone calls.

The network manager may choose to admit all requests under certain low-use circumstances. It is important to note that admissions control is a way to preserve the integrity of the calls and sessions that are in progress when a user requests access. Policies can also be implemented to terminate an ongoing call to process a higher-priority call request.

Call Accounting

Once the call is terminated, the MC notifies the accounting entity of the call details. The MC can also function in cooperation with back-office systems to generate billing. The information generated can include many details such as call duration, origin, destination, and QoS.

Bandwidth Management

The MC can control and limit the number of terminals allowed to use the network simultaneously. Through signaling, the MC is able to limit the bandwidth of the call to less than what was requested as well as reject calls from a terminal if it determines that there is not sufficient bandwidth available on the network to support the call. The MC can work in conjunction with a QoS server to achieve better QoS for calls. This function also operates during an active call when a terminal user requests additional bandwidth.

Call Management Services

The MC is able to maintain a list of ongoing calls that is similar to PBX logs. This information is necessary to indicate that a called terminal is busy and to provide information for the bandwidth management function.

Supplementary Services

Supplementary services per the standard, such as call forward and transfer, are critical telephony functions enterprise users will expect their network to provide. Both the MC and terminal can provide support for supplementary services; however, the MC performs it with less computational complexity and load on the client.

Directory Services

The MC's database contains user profiles that provide the necessary information for implementing directory services to help users find one another. It can access other directory services (such as Internet locator service [ILS]) that are updated or configured with the necessary information for connecting calls.

The MC interfaces to the other devices on the IP network using any one of a number of different protocols depending on the product developer and the needs of a particular implementation. The MC controls media gateways through protocols such as H.323, MGCP, or Megaco/H.248, Signaling Gateways are typically controlled through the SIGTRAN protocol.

These protocols will be discussed in detail in later chapters.

Multipoint Controller Units

MCUs provide support for conferences of three or more terminals or gateways. All terminals participating in the conference establish a connection with the MCU. The MCU manages conference resources, negotiates between terminals for the purpose of determining the audio or video coder/decoder (CODEC) to use, and may handle the media stream. MCUs may control any one or more of three different conference types.

Centralized
All parties on the conference communicate to the MCU on a point-to-point basis. The MCU manages the conference and also receives, processes, and sends the communications streams to and from the participating terminals.

Decentralized
In a decentralized conference the MCU is not directly involved in the conference. The end terminals communicate directly with each other. In this case the terminals assume the responsibility for combining the communications streams and maintaining the connections. This is typically done using IP multicast sessions.

Mixed
As the name implies, this conference method involves both centralized and decentralized modes. In this case the MCU connects all of the centralized parties to the decentralized conference and maintains that connectivity.

The MCs, gateways, and MCUs are logically separate components but can be implemented as a single physical device.

Adjunct Processors

There are many specialized processors that can be used in an IP-based converged network. These devices provide specific services required for a given network implementation. These devices are also used to offload the MGC for certain feature functionality. Often the devices are referred to as "back-end" devices.

There are many different types of these servers available in the industry. Here we will discuss some of the more common ones.

Call Accounting

Call accounting servers work in conjunction with the call accounting features on the media controller. This device serves the same purpose as the Station Message Detail Recording (SMDR) features on a PBX. Often the same call accounting adjuncts used in a traditional PBX environment have optional configurations to be used in an IP-based network.

The call accounting server collects from the MC all of the details regarding items such as called and calling parties, call duration, and any special events or codes that were generated during the call. The primary use of the information gathered in this server is to track calling patterns, determine if any additional resources are needed, and most commonly to bill back services to end users and organizations.

Systems Management

The systems management server is a device used by the network administrator to manage the multimedia applications on the converged network. This server often works in conjunction with, and is sometimes integrated into, the IP network management tools, such as IP switch and router management and QoS management. There are three major functions of the systems management server.

First, the systems management server is used to administer the converged network devices such as terminal configuration on the IP network and configuration of the trunks and stations associated with all of the media and signaling gateways. This is where all of the permissions, capabilities, and services that are given to a particular endpoint or gateway are determined. This is also the tool used to perform moves, adds, and changes of media controller, terminals and gateways in the network.

The second major function of the systems management server is to perform monitoring of the converged network. This server often has the ability to monitor the operation of the various components in the network and also can monitor the traffic patterns in the network. As part of the monitoring function, many systems management servers have a facility to notify a device, administrator, or user if any of the monitored functions are not working within optimal parameters.

The third main function is to perform systems maintenance and repairs. Based on the information gathered in monitoring the network, this server can be used to reset or reprogram the elements on the network. The server is also used in many implementations to support upgrades to the system.

Advanced Call Processing

The term Advanced Call Processing server refers to any server that adds additional multimedia calling functionality beyond the basic call processing features that are integrated into the MGC. The most common example of this type of server is a Contact Center or Customer Relationship Management (CRM) server. In a standard contact center implementation on a converged IP network, the more complex functions of the contact center are often off-loaded onto a specialized server.

In an inbound contact center application, when a call is presented to the media gateway from outside the network, the media gateway collects the information from the call (called party number, number dialed, etc.) and passes that information to the MC for instructions on how to connect the call. The MC then can send that information to the Contact Center processor and the contact center can use its internal algorithms to determine what endpoint should receive that call and in turn pass that information to the MC, which then instructs the media gateway as to where on the network to route the call.

This arrangement allows such processor-intensive and highly specialized functions as most-idle-agent, agent skill-based routing, load balancing, and other contact center specific applications to be done by a dedicated processor; the MC and media gateway only need to connect calls across the network as normal. Offloading these types of features onto specialized processors also allows media gateways and MCs to be used in a wide variety of networking applications without the need of overspecialization.

Computer/Telephony Interface

A Computer/Telephony Interface (CTI) is a device on the network that links the telephony applications of the MC and any advanced call processors to other servers in the network. These connections are used to access items such as account records and other sources of business intelligence to help the advanced processor and the MC make routing decisions.

Directory Access

The Directory Access server provides for access to outside directory services from the media controller and/or the advanced applications server. Usually this is done using the Lightwight Directory Access Protocol (LDAP) and the network to use external directory information for systems management and call control purposes.

Terminals

A terminal is the end-user communications device that is used for real-time bidirectional multimedia communications. A terminal can either be a personal computer (PC) or a stand-alone device. A vast array of terminal devices is available. Some of the most common types are

- a PC running a multimedia application such as a software-based voice telephone (softphone) or a streaming video conferencing application.
- a Desktop or Conference Room video unit.
- an H.323 or SIP-based telephone.

We will discuss the most common IP endpoint types in a later chapter.

There are also specialized types of terminals used for specific purposes on the converged network.

Messaging

Messaging Servers are specialized servers that provide a means for collecting, storing, and retrieving multimedia messages for the network terminal users. These servers are often set up on the network as a set of specialized terminals in the network.

An example of the use of a messaging server is for call answering purposes. In this case, a call is placed from one terminal to another. Let's call the terminal that starts the call the calling terminal and the intended recipient of the call the called terminal. The MC will instruct the calling terminal or its associated media gateway as to how to route the call to the other party. In some cases, if the call is not answered by the called terminal or the called terminal has invoked a do-not-disturb feature, the MC will tell the calling terminal or its associated media gateway to re-route the call to the terminal address associated with a voice messaging server. Once the message is taken, the messaging server can instruct the MGC to update the status on the called party terminal to light a "Message Waiting" indicator.

Integrated Voice Response

An Integrated Voice Response unit (IVR) is another specialized type of terminal, used to terminate an incoming call and provide an automated interface to the caller. IVRs are used in one of two main functions.

The first is to provide an automated information source such as a pre-recorded announcement or to provide automated service applications and other fixed-response type applications. The other major application on an IVR is to use it in conjunction with call routing functions to gain more information on a call. The feature is often used to prompt callers for identifying information. That information can then be sent to the call routing devices (the MC or a contact center server) to further help determine how to route the call.

Call Recording

The call recording server is another specialized terminal that is used to facilitate the recording of calls. Call recording servers are often used as terminals and are connected into a call as a third party either using an MCU or through multicast packets. Call recording is most often found in conjunction with contact centers and in financial transactions.

Throughout this chapter we have given descriptions of the major components of a theoretical converged, voice/data network. We have defined terms such as media gateway, media controller, terminal, and others. These terms refer to these units as logical elements of the network, and the functionality of the elements have been described.

In following chapters we will take these theoretical network elements and describe how they have been implemented in real, commercial products and applications.

CHAPTER 4

VoIP TECHNOLOGY

In this chapter, we will discuss the unique issues that need to be considered when deploying voice across an IP-based network infrastructure. We will discuss the special needs that voice traffic has, such as minimal delay, jitter, and packet loss. Also, we will discuss the network protocols that can help support those needs, including 802.1p. Diffserv and Priority Queuing. Finally we will provide some insight into some recommendations for the design of a converged network infrastructure.

Voice Traffic Needs in an IP Network

Voice quality is always a subjective topic. Defining "good" voice quality varies with business needs, cultural differences, customer expectations, and hardware/software. The final determination of acceptable voice quality lies within the user's definition of quality and design, implementation and monitoring of the end-to-end data network.

Quality is not one discrete value where the low side is good and the high side is bad. There is a tradeoff between real-world limits and acceptable voice quality. Lower delay, jitter, and packet loss values can produce the best voice quality, but also may come with a cost to upgrade the network infrastructure to achieve the low network values. For example, in a WAN link between Los Angeles and Manila, the link could add a large, fixed delay into the overall delay budget. This delay in the conversation may be noticeable to the end users as either an actual delay, or as echo, so the users may not get "toll quality" voice, but the cost savings can far override the slight loss of quality and the voice quality can still be acceptable for the user's purposes.

The concept of voice quality is also greatly influenced by the user's business needs. IP telephony quality can be engineered to several different levels to accommodate differing business needs. A small company may choose to implement IP telephony with acceptable sound instead of buying newer networking equipment to support excellent voice sound. A large call center company may want excellent voice sound as part of its corporate strategy, regardless of the cost of implementation.

The voice network has traditionally been separate from the data network because of the protocols used as well as the fact that the characteristics of voice applications are very different from those of data applications. Voice calls have had their own dedicated bandwidth throughout the circuit-switched network. This provided an environment where extremely high levels of reliability became the standard. Interactive voice traffic is sensitive to delay and jitter but can tolerate some packet loss, problems that were rarely an issue with circuit switching.

The data network, on the other hand, is packet switched. Data is less sensitive to delay and jitter, but cannot tolerate loss. The data philosophy has centered on providing reliable data transmission over unreliable media, almost regardless of delay. Bandwidth in the data world is largely shared, so congestion and delay are often present and can cause problems for multimedia applications such as voice.

The factors that affect the quality of data transmission are different from those affecting the quality of voice transmission. Data is generally not greatly affected by delay. Voice transmissions are degraded by relatively small amounts of delay and cannot be retransmitted. Additionally, a small amount of packet (data) loss does not affect voice quality at the receiver's ear, but even a small loss of data can corrupt an entire file or application. So in some cases, introducing VoIP to a high-performing data network can yield very poor voice quality.

Implementing VoIP requires attention to many factors, including

- Delay
- Jitter
- Packet loss
- Packet mis-order
- Available bandwidth
- Packet prioritization
- Network design
- Endpoint audio characteristics (sound card, microphone, earpiece, etc.)
- Transcoding
- Echo
- Silence suppression
- Duplex
- Codec selection
- Router and data-switch configuration
- WAN protocols

- QoS/CoS policy
- Encryption/Decryption

Delay

Packet delay is the length of time it takes a packet to traverse the network. Each element of the network—switches, routers, distance traveled through the network, firewalls and jitter buffers (such as those built into H.323 audio applications like Microsoft NetMeeting)—adds to packet delay. Router delay depends not only on hardware, but also on configurations such as access lists, queuing methods, and transmission modes. Delay (latency) can have a noticeable effect but can be controlled somewhat in a private environment (LAN/WAN) because the company or enterprise manages the network infrastructure or SLA. When using the public network, there are inherent delays that one cannot control.

VoIP network vendors give many differing guidelines as to the amount of delay that is acceptable to maintain good voice quality. A good set of guidelines for voice quality is

- Under 150–200 ms delay can give very good voice quality.
- Delays exceeding 200 ms may still be quite acceptable depending on customer expectations, analog trunks used, codec type, etc.
- The H.323 protocol defines a maximum end-to-end delay of 400 ms. Delays beyond this level can cause network instability.
  (These numbers are for delay between endpoints, meaning LAN/WAN measurements not including IP phones.)

The ITU-T has recommended 150 ms one-way delay (including endpoints) as the limit for "excellent" voice quality. One-way delays in excess of 250 ms can cause "talk-over," which is when each person starts to talk because the delay prevents them from realizing that the other person has already started talking.

Long WAN transports must be considered as a major contributor to the network delay budget. Some WAN service providers can lower delay in their network if it is negotiated and recorded as part of the company's SLA (Service Level Agreement).

Jitter

Because network congestion can be encountered at any time within a network, buffers can fill instantaneously. This instantaneous buffer utilization can lead to a difference in delay times between packets in the same voice stream. This difference, called jitter, is the variation between when a packet is expected to arrive and when it actually is received.

Jitter can create audible voice-quality problems if the variation is greater than 20 ms (assuming an existing 20 ms packet size). Symptoms of excessive jitter are very similar to symptoms of high delay, because in both cases packets are discarded if the packet delay exceeds half the jitter buffer size.

To compensate for network jitter, many vendors implement a jitter buffer in their H.323 voice applications. The purpose of the jitter buffer is to hold incoming packets for a specified period of time before forwarding them to the decompression process. A jitter buffer is designed to smooth packet flow. In doing so, it can also add packet delay. Jitter buffers should be dynamic to give the best quality, or if static, should generally be sized to twice the largest statistical variance between packets. Router vendors have many queuing methods that alter the behavior of the jitter buffer. It is not enough to just select the right size of jitter buffer, one must also pair an appropriate queue-unloading algorithm type with the jitter buffer. The network topology can also affect jitter. Because there are fewer collisions on a data-switched network than on a hub-based network, there will be less jitter on the switched network.

Packet Loss

Network packet loss occurs when packets are sent, but not received at the final destination due to some network problem. Qualifying problems caused by occasional packet loss are difficult to detect because each codec has its own packet loss concealment method. Therefore, it is possible that voice quality would be better using a compression codec (G.729A) compared to a full bandwidth G.711 codec. Several factors make packet loss requirements somewhat variable, such as the following:

- Packet loss requirements are tighter for tones (other than DTMF) than for voice. The ear is less able to detect packet loss during speech (variable pitch), than during a tone (consistent pitch).
- Packet loss requirements are tighter for short, continuous packet loss than for random packet loss over time. Losing ten contiguous packets is worse than losing ten packets evenly spaced over an hour timespan.
- Packet loss may be more noticeable for larger voice payloads than for smaller ones, because more voice is lost in a larger payload.
- Packet loss may be more tolerable for one codec over another.
- Even small amounts of packet loss can greatly affect TTY (TDD) devices' ability to work properly.
- Packet loss for TCP signaling traffic increases substantially over 3 percent loss due to retransmissions.

Again, acceptable rates for packet loss vary with the needs of the end users.

- 1 percent or less can yield toll quality depending on many factors.
- 3 percent or less should give business communications quality.

- More than 3 percent may be acceptable for voice but may interfere with signaling.

Packet Mis-Order

Network packet mis-order is, for VoIP, very much like packet loss. If a packet arrives out of order, it is generally discarded as it makes no sense to play it out of order and buffers are small. Specifically, packets are discarded when they arrive later than the jitter buffer can hold them. Mis-order can occur when networks send individual packets over different routes. Planned events such as load balancing, unplanned events such as re-routing due to congestion, or other transient difficulties can cause packet mis-order. Packets traversing the network over different routes may arrive at their destination out of order. Network latency over multiple yet unequal routing paths can also force packet mis-order.

Transcoding

Transcoding is a voice signal converted from analog to digital or digital to analog (possibly with or without compression and decompression). If calls are routed using multiple voice coders, as in the case of call coverage on an intermediary system back to a centralized voice mail system, the calls may experience multiple transcoding (including the one in and out of the voice mailbox). Each transcoding episode results in some degradation of voice quality.

These problems may be minimized by the use of some form of network rerouting (e.g., Q-SIG Path Replacement). This feature detects that the call coming through the main call server has been routed from one tandem node, through the main, and back out to a third switch. In these cases, the system then re-routes the call directly, thus replacing the path through the main system with a more direct connection.

Echo

The two main types of echo are acoustic and impedance, although the sources of echo can be many. Echo will result when a VoIP call leaves the LAN through a poorly administered analog trunk into the PSTN. Another major cause is from an impedance mismatch between four-wire and two-wire systems. Echo also results when an impedance mismatch exists in the conversion between TDM (Time Division Multiplexing) networks and the LAN, or the impedance mismatch between a headset and its adapter. Impedance mismatch causes inefficient energy transfer. The energy imbalance must go somewhere and so it is reflected back in the form of an echo. Usually the speaker hears an echo but the receiver does not.

Echo cancellers, which have varying amounts of memory, compare the received voice with the current voice patterns. If the patterns match, the canceller cancels the echo. Echo cancellers aren't perfect, however. Under some circumstances, the echo gets past the canceller. The problem is exacerbated in VoIP systems. If the one-way trip delay between endpoints is larger than the echo canceller memory, the echo canceller won't ever find a pattern to cancel.

Silence Suppression

Silence suppression software, often called Voice Activity Detection (VAD), monitors the received signal for voice activity. When no activity is detected for the configured period of time, the software informs the Packet Voice Protocol. This prevents the encoder output from being transported across the network when there is silence, resulting in additional bandwidth savings. The software also measures the idle noise characteristics of the telephony interface. It reports this information to the Packet Voice Protocol to relay this information to the remote end for comfort noise generation when no voice is present.

Some VADs can cause voice clipping and can result in poor voice quality, but the use of VAD can greatly conserve bandwidth and is therefore a very important detail to consider when planning network bandwidth—especially in the WAN (Wide Area Network).

Duplex

The ideal LAN network for transporting VoIP traffic is a network that is fully switched from end to end because it significantly reduces or eliminates collisions. A network that has shared segments (hub-based) can result in lower voice quality due to excessive collisions and should be avoided.

Codec Selection

Depending on the bandwidth availability and acceptable voice quality, it might be worthwhile to select a codec that produces compressed audio.

- A G.711 codec produces audio uncompressed to 64 Kbps
- A G.729 codec produces audio compressed to 8 Kbps
- A G.723 codec produces audio compressed to 5.3 to 6.3 Kbps

Table 4.1 provides comparisons of several voice quality considerations associated with some of the codecs. These qualities are based on an industry guideline know as the Mean Opinion Score (MOS). The MOS is a subjective scale

Standard	Coding Type	Bit Rate (Kbps)	MOS
G.711	PCM	64	4.3
G.729	CS-ACELP	8	4.0
G.723.1	ACELP	6.3	3.8
	MP-MLQ	5.3	

Table 4.1

Comparison of common speech coding standards

for rating voice quality from 1 to 5, where 4 or better is considered "toll quality" or the equivalent of the quality in the U.S. public switched telephone network (PSTN).

Generally, G.711 is used within LANs because bandwidth is abundant and inexpensive, whereas G.729 is used across WAN links because of the bandwidth savings and reliable voice quality.

PC Considerations Using IP SoftPhone

IP SoftPhones are software on a PC that simulates a telephone. The "perceived" audio/voice quality at the PC endpoint is a function of at least four factors, as follows:

1. Transducer Quality
   The selection of speaker and microphone or headset has an impact on the reproduction of the sound.
2. Sound Card Quality
   There are several parameters that affect sound card quality. The most important is whether or not the sound card supports full-duplex operation.
3. End-to-End Delay
   A PC can be a major component of delay in a conversation. PC delay consists of the jitter buffer and sound system delays, as well as the number of other processes running and the speed of the processor.
4. Speech Breakup
   Speech breakup may be the result of a number of factors.
   - Network jitter in excess of the jitter buffer size
   - Loss of packets (due to excessive delay, etc.)
   - Aggressiveness of silence suppression
   - Performance bottleneck in the PC

Lower-speed PCs (or PCs with slow hard drives) may have adverse interactions with sound playback and recording. This can cause breaks in received or transmitted audio. The best thing to do in this situation is to increase the processor speed, increase the amount of RAM, and/or reduce the number of applications competing for the processor or hard drive resources. One notable resource consumer is the Microsoft Find Fast program that launches from the Startup folder (and runs in the background). This application periodically re-indexes the hard drive and consumes significant PC resources in the process.

Bandwidth Requirements

The bandwidth available to the user is very important. Access to the network using slower connections, such as dial-up connections, will degrade voice quality. The best voice quality is achieved in both LANs and WANs when the bandwidth is "owned" by the customer. Customer-owned bandwidth can be shaped to optimize VoIP traffic. Conversely, bandwidth that is not controlled, like the Internet, cannot give consistent sound quality because it cannot be optimized for VoIP. Factors of delay, jitter, and packet loss are exacerbated over the public Internet, making the assurance of voice quality over the internet especially problematic.

The bandwidth required for a given call varies with the voice codec used. Each codec converts the voice stream into a digital representation using varying amounts of bits to represent a second of voice. In addition to the codec's bit rate, the IP packet framing adds overhead to the data stream.

Properly provisioning the network bandwidth is a major component of designing a successful network. You can calculate the required bandwidth adding the bandwidth requirements for each major application (for example, voice, video, and data). This sum then represents the minimum bandwidth requirement for any given link. Many VoIP manufacturers recommend that minimum bandwidth should not exceed approximately 75 percent of the total available bandwidth for the link. This 75 percent rule assumes that some bandwidth is required for overhead traffic, such as routing and Layer 2 keep-alives, as well as for additional applications such as e-mail and Hypertext Transfer Protocol (HTTP) traffic.

As illustrated in Table 4.2, a VoIP packet consists of the payload, IP header, User Datagram Protocol (UDP) header, Real-time Transport Protocol (RTP) header, and Layer 2 Link header. At a packetization rate of 20 ms, VoIP packets have a 160-byte payload for G.711 or a 20-byte payload for G.729. The IP header is 40 bytes, the UDP header is 8 bytes, and the RTP header is 12 bytes. The link header varies in size according to media.

The bandwidth consumed by VoIP streams is calculated by adding the packet payload and all headers (in bits), then multiplying by the packet rate per second.

Voice Payload	RTP Header	UDP Header	IP Header	Link Header
X Bytes	12 Bytes	8 Bytes	20 Bytes	X Bytes

Table 4.2

Typical VoIP packet

This does not include Layer 2 header overhead and does not take into account any possible compression schemes, such as compressed Real-time Transport Protocol (cRTP).

Table 4.3 shows the typical bandwidth need for common codecs and sampling rates.

CODEC	Sampling Rate	Voice Payload in Bytes	Bandwidth per Conversation
G.711	20 ms	160	80 Kbps
G.711	30 ms	240	53 Kbps
G.729A	20 ms	20	24 Kbps
G.729A	30 ms	30	16 Kbps

Table 4.3

Bandwidth consumption for voice payload only

Table 4.4 shows the total bandwidth, including layer 2 headers for various transmission media.

CODEC	Ethernet 14 Bytes of Header	PPP 6 Bytes of Header	ATM 53-Byte Cells with a 48-Byte Payload	Frame Relay 4 Bytes of Header
G.711 (20 ms sample rate)	85.6 Kbps	82.4 Kbps	106 Kbps	81.6 Kbps
G.711 (30 ms)	56.5 Kbps	54.4 Kbps	70 Kbps	54 Kbps
G.729A (20 ms)	29.6 Kbps	26.4 Kbps	42.4 Kbps	25.6 Kbps
G.729A (30 ms)	19.5 Kbps	17.4 Kbps	28 Kbps	17 Kbps

Table 4.4

Bandwidth consumption with headers included

These numbers are for each direction in a conversation, so a two-way conversation needs double this bandwidth across a given link.

There are methods to lessen some of these bandwidth needs. One is RTP header compression (cRTP), which reduces the RTP header from 12 bytes per frame to 4. Another is to use silence suppression to suppress the packets sent by the non-speaking party in a conversation.

IP Network Protocols to Support Voice

Voice quality is only as good as the quality of the weakest network link. Packet loss, delay, and delay variation all contribute to degraded voice quality. In addition, because network congestion (or more accurately, instantaneous buffer congestion) can occur at any time in any portion of the network, network quality is an end-to-end design issue. The QoS tools discussed here are a set of mechanisms to increase voice quality on data networks by decreasing dropped voice packets during times of network congestion and by minimizing both the fixed and variable delays encountered in a given voice connection.

In Chapter 14, we will discuss all of these QoS schemes in more detail. For our purposes here we will discuss how these QoS mechanisms can be used to optimize voice quality in a converged network.

In order for a VoIP solution to function well, the network must be able to give voice packets priority over ordinary data packets or sufficient bandwidth must always be available. There are several strategies available to provide this prioritization. These strategies include using class of service (CoS), prioritizing ports, prioritizing services, and using IEEE 802.1 p/Q to set the priority bits.

CoS versus QoS

The concepts of Class of Service (CoS) and Quality of Service (QoS) are often confused. Class of Service is a classification method only. CoS does NOT ensure a level of Quality of Service, but is the method used by queuing mechanisms to limit delay and other factors to improve QoS. Most CoS strategies assign a priority level, usually 0–7 or 0–63, to a frame or packet respectively. Common CoS models include the IP TOS (Type Of Service) byte, Differentiated Services Code Point (DiffServ or DSCP, defined in RFC 2474 and others), and the IEEE 802.1 p/Q.

Quality of Service (QoS) involves giving preferential treatment through queuing, bandwidth reservation, or other methods based on attributes of the packet, such as CoS priority. A service quality is then negotiated. Examples of QoS are

CBWFQ (Class-Based Weighted Fair Queuing), RSVP (RESERVATION Protocol—RFC 2205), MPLS (Multi Protocol Label Switching—RFC 1117 and others).

CoS, or tagging, is totally ineffective in the absence of QoS because it can only mark data. QoS relies on tags or filters to give priority to data streams.

UDP Ports

One prioritization scheme assigns priority based on the UDP (User Datagram Protocol) port numbers used by the voice packets. This scheme allows the use of network equipment to prioritize all packets from a port range. UDP is used to transport voice through the LAN because, unlike TCP, it is not connection based. Because of the human ear's sensitivity to delay, it is better to drop packets rather than retransmit voice in a real-time environment, so a connectionless protocol is preferable to a connection-based protocol.

VoIP devices and servers can define a port range for voice priority. Routers and layer 3 data switches can then use these ports to distinguish priority traffic. This priority traffic can be voice packets (UDP), signaling packets (TCP), or both. This is an OSI model layer 4 solution and works on data coming to and from the specified ports or port range.

Differential Services (DiffServ, DSCP, TOS)

The Differential Services Code Point (DSCP) prioritization scheme redefines the existing Type of Service (ToS) byte in the IP header by combining the first 6 bits into 64 possible combinations. This use of the ToS byte is still evolving but can be used now by vop media controllers, IP telephones, and other network elements such as routers and switches in the LAN and WAN.

Please note that older routers may require a DSCP setting of 40 (101,000), which is backward compatible to the original ToS byte definition of critical. The ToS byte is an OSI model layer 3 solution and works on IP packets on the LAN and possibly the WAN depending on the service provider.

IEEE 802.1 p/Q

Yet another prioritization scheme is the IEEE 802.1Q standard, which uses four bytes to augment the layer-2 header. IEEE 802.1Q defines the open standard for VLAN tagging. Two bytes house twelve bits used to tag each frame with a VLAN identification number. The IEEE 802.1p standard uses three of the remaining bits in the 802.1Q header to assign one of eight different classes of service. IEEE 802.1Q are OSI layer-2 solutions and work on frames.

VLANs

VLANs provide limited security and create smaller broadcast domains through software by creating virtually separated subnets. Broadcasts are a natural occurrence in data networks from protocols used by PCs, servers, switches, routers, NOS, etc. Creating a separate VLAN for voice reduces the amount of broadcast traffic (and unicast traffic on a shared LAN) the telephone will receive. Separate VLANs result in more effective bandwidth utilization and reduce the processor burden on the IP telephones and PCs by freeing them from having to analyze irrelevant broadcast packets. VLANs, a layer-2 feature, are created in data switches. A voice VLAN can also be manually applied to an IP telephone or given by a DHCP server. CoS tagging and QoS policies can be applied at OSI layer 2 by using VLANs.

Separate voice and data VLANs are options that make sense for many users. Note, however, that VLAN implementation and maintenance can be substantial, and again, is an option even as a best practice.

Resource Reservation Protocol

Once a given packet has been labeled by one or more of the schemes already mentioned, there are two common methods of acting on those labels to give the voice packets priority. One method is to act on a per-packet basis by giving special treatment to the packets as they are processed. The second is to establish a specialized route for those packets to traverse in advance.

In this section we will talk about the most common protocol used to support that second method, the Resource Reservation Protocol (RSVP).

RSVP is a signaled QoS configuration mechanism. It is a protocol by which applications can request end-to-end, per-conversation, QoS from the network and can indicate QoS requirements and capabilities to peer applications. RSVP is a layer-3 protocol suited primarily for use with IP traffic.

There are two significant RSVP messages, PATH and RESV. Transmitting applications send PATH messages toward receivers. These messages describe the data that will be transmitted and follow the path that the data will take. Receivers send RESV messages. These follow the path seeded by the PATH messages, back toward the senders, indicating the profile of traffic that particular receivers are interested in. In the case of multicast traffic flows, RESV messages from multiple receivers are "merged," making RSVP suitable for QoS with multicast traffic.

As defined today, RSVP messages carry the following information:

- How the network can identify traffic on a conversation (classification informaton)

- Quantitative parameters describing the traffic on the conversation (data rate, etc.)
- The service type required from the network for the conversation's traffic
- Policy information (identifying the user requesting resources for the traffic and the application to which it corresponds).

How RSVP Works

PATH messages wind their way through all network devices en route from sender to receivers. RSVP-aware devices in the data path note the messages and establish state for the flow described by the message.

When a PATH message arrives at a receiver, the receiver responds with a RESV message (if the receiving application is interested in the traffic flow offered by the sender). The RESV message winds its way back toward the sender, following the path established by the incident PATH messages. As the RESV message progresses toward the sender, RSVP-aware devices verify that they have the resources necessary to meet the QoS requirements requested. If a device can accommodate the resource request, it installs classification state corresponding to the conversation and allocates resources for the conversation. The device then allows the RESV message to progress on toward the sender. If a device cannot accommodate the resource request, the RESV message is rejected and a rejection is sent back to the receiver.

In addition, RSVP-aware devices in the data path may extract policy information from PATH messages and/or RESV messages, for verification against network policies. Devices may reject resource requests based on the results of these policy checks by preventing the message from continuing on its path and sending a rejection message.

When requests are not rejected for either resource availability or policy reasons, the incident PATH message is carried from sender to receiver and a RESV message is carried in return. In this case, a *reservation* is said to be installed. An installed reservation indicates that RSVP-aware devices in the traffic path have committed the requested resources to the appropriate flow and are prepared to allocate these resources to traffic belonging to the flow. This process of approving or rejecting RSVP messages is known as *admission-control* and is a key QoS concept.

Guarantees must be valid *end-to-end* across multiple subnets. Lower quality guarantees can be provided without requiring tight coupling between the QoS mechanisms in different subnets. However, high quality guarantees require tight coupling between these mechanisms.

As an example, it is possible to independently configure devices in each subnet (in a top-down manner) to prioritize some set of traffic (as identified by IP port) above best-effort traffic (BBE service). This will indeed improve the quality of service perceived by the prioritized application, in all parts of the network. However, this is a low quality guarantee as it makes no specific commitments regarding available bandwidth or latency.

On the other hand, consider the quality of guarantee required to support a videoconference. A videoconferencing application requires that all subnets between the videoconferencing peers be able to provide a significant amount of bandwidth at a low latency. To do so efficiently requires that all devices along the data path commit the required amount of low latency bandwidth for the duration of the videoconference. As we have seen, high quality guarantees such as these generally require signaling across network devices in order to make efficient use of network resources. In our sample network, multiple subnets, based on multiple media (and varying traffic handling mechanisms) must be coordinated via this signaling. RSVP is particularly suitable for this purpose because it expresses QoS requirements in high-level, abstract terms. Agents in each subnet are able to translate the media-independent, abstract requests into parameters that are meaningful to the specific subnet media.

Hosts generate RSVP signaling when it is necessary to obtain high quality guarantees. The network listens to this signaling at strategic points in the network. We will refer to devices that participate in RSVP signaling as *RSVP agents* or alternatively as *signaling* or *admission control agents*. As we have shown, appointing such agents at varying densities can provide varying quality/ efficiency products. At a minimum we assume one or more admission control agents in each subnet. Each agent uses the mappings defined in ISSLL to translate high-level end-to-end RSVP requests into parameters that are meaningful to the media for which the agent is responsible. The admission control agent then determines, based on resource availability and/or policy decisions (with the cooperation of PDPs), whether an RSVP request is admissible or not. Any admission control agent along the route from sender to receiver may veto an RSVP request for resources. Requests that are not vetoed by any device are considered admitted and result in the return of an RSVP RESV message to the requesting transmitting host.

Prioritized Queuing

As we discussed in the previous section, another common QoS mechanism is to give special treatment to packets marked as voice packets. This special treatment involves some form of prioritized queuing.

Prioritized queuing is a generic term for one of a number of schemes that allow one type of traffic to have priority over another when being transmitted over a bandwidth-limited media. The basic idea is for the queuing device (usually an IP switch or router) to analyze the packets, determine the type of payload the packet is carrying, compare that information to a set of prioritization and bandwidth utilization rules, and order the output according to those rules.

There are a number of different prioritization schemes used in currently available network devices today. The most commonly used schemes are known as Priority Queuing, Weighted Fair Queuing, and Class-Based Queuing. Many vendors use different names for these schemes and often combine the schemes together.

Priority Queuing

In Priority Queuing, packets are forwarded on a strict priority basis. This means that packets in the higher-priority queue are sent first before any packets in the lower-priority queue (see Figure 4.1).

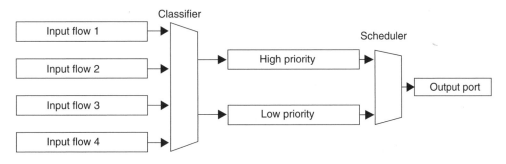

Figure 4.1

Priority queuing.

This works well for giving highly predictable performance for the high-priority packets, but lower-priority packets can be greatly delayed if there is significant high-priority traffic. Priority queuing works best in situations where the high-priority traffic has a relatively fixed bandwidth requirement (like voice). That allows priority queuing to be use while still giving predictable performance to the lower-priority traffic.

Weighted Fair Queuing

Weighted Fair Queuing (WFQ) is a method to overcome problems in strict Priority Queuing by giving all of the packets a weighting value based on priority and the anticipated time a given packet needs for processing. Once the packets are all given weights, they are transmitted in weight order. This gives the ability to still favor high-priority traffic, but give a fairer distribution of bandwidth to all other traffic types (see Figure 4.2).

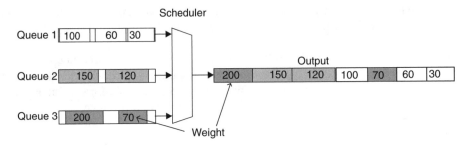

Figure 4.2

Weighted fair queuing.

The biggest problem with WFQ is that it is a very complex mechanism and therefore must be implemented in software, so the WFQ process can add delay to the overall transmission.

Class-Based Queuing

Class-Based Queuing (CBQ) is a form of queuing where incoming packets are sorted into classes as in the other queuing schemes, each class is queued separately and assigned a fixed percentage of the output bandwidth. The output of the queues is then delivered in a given time period based on its percent allocation (see Figure 4.3).

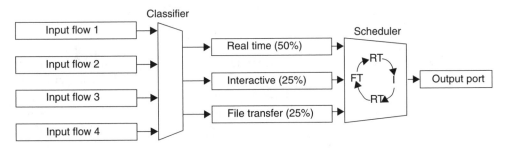

Figure 4.3

Class-based queuing.

This scheme allows all service types to have predictable performance. Also, due to the simpler nature of the queuing mechanisms, CBQ can be done in hardware, allowing for greater overall performance.

Most commercially available, converged network devices use one or more of these schemes to support simultaneous voice, data, and video traffic. The vast majority use either a class-based queuing scheme or a hybrid scheme where fixed bandwidth, high-priority traffic like voice or video conferencing are queued in a priority queue while the rest of the IP services use some form or WFQ or CBQ at the second level.

Other Elements that Affect VoIP

WAN Considerations

Until WAN bandwidth becomes affordable at any speed, delivering bandwidth to applications over the WAN will remain a formidable task. When voice traffic is carried on packet networks, different labeling or queuing schemes function to give voice packets priority over data packets. The presence of large data packets may result in added serialization delay for VoIP packets across WAN links. This is due to the fact that smaller VoIP packets are held in queue while larger data packets are processed onto the WAN link. To avoid excessive delay, there may be benefit to fragmenting the larger data packets and interleaving them with the smaller voice packets. One technique is to adjust the packets by adjusting the Maximum Transmission Unit (MTU) size. Other techniques, such as Multilink PPP (MPP) Link Fragmenting and Interleaving (LFI) and Frame Relay Fragmentation (FRF12), allow network managers to fragment larger packets as well as allow queuing mechanisms to speed the delivery of Real Time Protocol (RTP) traffic without significantly increasing protocol overhead or reducing data efficiency. Also, header compression protocols such as cRTP (Compressed Real Time Protocol) can and should be used between WAN links. Hardware-based cRTP is effective with very minimal delays, but software cRTP can add significant delay.

VPN (Virtual Private Network)

There are many definitions for Virtual Private Networks (VPN). For our purposes, VPNs refer to encrypted tunnels carrying packetized data between remote sites. VPNs can use private lines or use the Internet via one or more Internet Service Providers (ISP). VPNs are implemented in both dedicated hardware and software, but can also be integrated as an application to existing hardware and software packages. A common example of an integrated package is a firewall product that can provide a barrier against unauthorized intrusion as well as perform the security features needed for a VPN session.

The encryption process can take from less than 1 millisecond to 1 second or more, at each end. Obviously, VPNs can represent a significant source of delay and, therefore, negatively affect voice performance. Also, as most VPN traffic runs over the Internet and there is little control over QoS parameters for traffic crossing the Internet, voice quality may suffer due to excessive packet loss, delay, and jitter. Users may be able to negotiate a service-level agreement with the VPN provider to guarantee an acceptable level of service.

Frame Relay

The obstacle in running VoIP over frame relay involves the treatment of traffic within the Committed Information Rate (CIR) and outside of CIR, commonly termed the "burst range."

Traffic up to the CIR is guaranteed while traffic beyond the CIR typically is not. This is how frame relay is intended to work. CIR is a committed and reliable rate, whereas burst is a bonus when network conditions permit it without infringing on any user's CIR. For this reason, burst frames are marked Discard Eligible (DE) and are queued or discarded when network congestion exists. Although experience has shown that customers can achieve significant burst throughput, it is unreliable and unpredictable and not suitable for real-time applications such as VoIP.

Therefore, the objective is to prevent voice traffic from entering the burst range and being marked DE. One way to accomplish this is to prohibit bursting by shaping the traffic to the CIR and setting the excess burst size (Be—determines the burst range) to zero. However, this also prevents data traffic from using the burst range as well. Another possible alternative is to size the CIR above the peak voice traffic level, and then prioritize the voice traffic so that it is always delivered first. The underlying assumption here is that the network administrator has an expectation of peak voice traffic. By sizing the CIR to meet or exceed the peak voice traffic and then applying priority queuing on the interface so that VoIP is serviced first, we can intuitively assure that voice traffic will not enter the burst range.

The problem with the latter method, however, is that the actual queuing mechanism is not always intuitive. Even though the aggregate voice traffic throughput cannot exceed the CIR, it is possible that a voice packet could be sent in the burst range. The technical workings of this are beyond the scope of this document. But simply stated, it is possible that a voice packet would arrive right after many data packets have already been transmitted in the CIR range, such that the voice packet ends up in the burst range when the router processes it. However, the method is certainly worth trying.

One good piece of knowledge is that most IXCs convert the long-haul delivery of frame relay into ATM. That is, the frame relay PVC is converted to an ATM PVC at the first frame relay switch after leaving the customer's premise. It is not converted back to frame relay until the last frame relay switch before entering the customer's premise. This has significance because ATM has built-in Class of Service (CoS). A customer can contract with a carrier to convert the frame relay PVC into a Constant Bit Rate (CBR) ATM PVC. ATM CBR cells are delivered with lower latency and higher reliability.

As a final note, the reader should understand that under the best circumstances, frame relay is still inherently more susceptible to delay than ATM or TDM. Therefore, after applying the best possible queuing mechanism, one should still expect more delay over frame relay than would be present over ATM or TDM.

Network Address Translation (NAT)

VoIP does not work well with networks that use NAT (Network Address Translation) because most NAT implementations do not support H.323 protocols. The destination IP address is encapsulated in more than one header: the Q.931, H.225, and IP headers. NAT changes only the address in the IP header, resulting in a mismatch that prohibits the control of calls. Using a firewall to guard against intruders, it is often suggested, but the firewall should not provide NAT functions for VoIP packets unless it is Q.931 friendly.

Network Design Recommendations

This section is an amalgam of various converged network vendors' recommendations for best practices in designing a converged IP network.

A network should be designed with the following factors in mind:

- Reliability/redundancy
- Scalability
- Manageability
- Bandwidth.

Supporting voice adds even more considerations when designing a network. These additional considerations include

- Delay
- Jitter
- Loss
- Duplex.

Generally speaking, these concerns dictate a hierarchical network consisting of at most three layers: core, distribution, and access. Some smaller networks can collapse the functions of several layers into one device.

The core layer is the heart of the network. Its purpose is to forward packets as quickly as possible. It should be designed with high availability in mind. Generally, these high-availability features include redundant devices, redundant power supplies, redundant processors, and redundant links. In the current era, core interconnections increasingly use Gigabit Ethernet.

The distribution layer links the access layer with the core. It is here that QoS features and access lists are applied. Generally, Gigabit Ethernet connects to the core and either Gigabit Ethernet or 100base-TX/FX links connect the access layer. Redundancy is important at this layer, but not as important as in the core.

The access layer connects servers and workstations. Switches at this layer are smaller, usually 24–48 ports. Desktop computers and workstations are usually connected at 10 Mbps (or 100 Mbps) and servers are connected at 100 Mbps (or 1 Gbps). Limited redundancy is used. Some QoS and security features can be implemented here.

For VoIP to work well, WAN links should be properly sized with sufficient bandwidth for voice and data traffic. Each voice call uses between 6.3 Kbps and 80 Kbps, depending on the desired codec, quality, and header compression used. Interoffice bandwidth demands can be sized using traditional phone metrics such as average call volume, peak volume, and average call length.

Quality of Service also becomes increasingly important with WAN circuits. In this case, Quality of Service can be taken to mean classification and prioritization of voice traffic. Voice traffic should be given absolute priority through the WAN, and if links are not properly sized or queuing strategies are not properly implemented, it will become evident both with the quality and timeliness of voice and data traffic.

There are three technologies that work well with VoIP: ATM, Frame Relay, and point-to-point (PPP) circuits. These technologies all have good throughput, low latency, and low jitter. ATM has the added benefit of enhanced QoS. Frame Relay and PPP links are more economical but lack some of the traffic-shaping features of ATM.

Of the three technologies, Frame Relay is the most difficult WAN circuit to use with VoIP. Congestion in Frame Relay networks can cause frame loss, which can significantly degrade the quality of VoIP conversations. With Frame Relay, proper sizing of the CIR (committed information rate) is critical. In a Frame Relay network, any traffic exceeding the CIR is marked discard eligible, and will be discarded at the carrier's option if it experiences congestion in its switches. It is very important that voice packets not be dropped. Therefore, CIR should be sized to average traffic usage. Usually, 25 percent of peak bandwidth is sufficient. Also, Service Level Agreements (SLAs) should be established with the carrier that define maximum levels of delay and frame loss, and remediation should the agreed-to levels not be met.

Network management is another important area to consider when implementing VoIP. Because of the stringent requirements imposed by VoIP, it is critical to have an end-to-end view of the network and ways to implement QoS policies globally. Products such as HP OpenView Network Node Manager, Visibility, Concord NetHealth, and MRTG will help administrators maintain acceptable service. Should a company not have the resources to implement and maintain network management, outsource companies are springing up to assist with this need.

Common Issues

Common network designs that can severely impact network performance when using VoIP include the following:

- Using a flat, non-hierarchical network: This can results in bottlenecks as all traffic must flow across the uplinks (at maximum 1 Gbps) versus traversing switch fabric (up to 256 Gbps). The greater the number of small switches (layers), the greater the number of uplinks and the lower the bandwidth for an individual connection. Under a network of this type, voice performance can quickly degrade to an unacceptable level.
- Multiple subnets on a VLAN: A network of this type can have issues with broadcasts, multicasts, and routing protocol updates. It can greatly impact voice performance and complicate troubleshooting issues.
- Hub-based network: Hubs in a network create some challenges for administrators. The collision domain, the number of ports connected by hubs without a switch or router in between, should be kept as low as possible. The effective (half-duplex) bandwidth available on a shared collision domain is approximately 35 percent of the total bandwidth available.
- Too many access lists: Access lists slow down a router. While they are appropriate for voice networks, care must be taken not to apply them to unnecessary interfaces. A good practice is to model the traffic beforehand, with access lists applied only to the appropriate interface in the appropriate direction, not all interfaces in all directions.

Additional concerns when implementing VoIP include:

- Network Address Translation (NAT): Due to limitations in the H.323 VoIP standard, VoIP conversations rarely work across NAT boundaries. It is important to route voice streams around routers or firewalls running NAT or use a H.323-friendly NAT.
- Virtual Private Networks (VPN): The encryption used with VPNs can add significant latency to voice streams, adversely affecting the voice quality.

Also, VPNs generally run over t\he Internet. Because there is no control over QoS parameters for traffic crossing the Internet, voice quality may suffer due to excessive packet loss, delay, and jitter.

In this chapter, we have discussed the special issues involved in the transmission of voice, a real-time application, over an IP network, a non-real-time media. We discussed the needs of the voice packets and how those needs can affect the quality of the voice transmission. We also discussed the various protocols and methodologies used to provide for good-quality voice in an IP network. The last section of the chapter also provided some recommendations for the design and implementation of a converged VoIP network.

Chapter 5

APPROACHES TO IMPLEMENTING VoIP ARCHITECTURES

In the preceding chapters, we discussed the components and underlying technological requirements of a theoretical VoIP-based converged network. In this chapter we will discuss the methods used by the various converged network providers to develop real-world implementations of these networks. We will discuss the most common methods of developing VoIP architectures and the relative strengths and weaknesses of these approaches. This will provide some insight into how the theories for converged network design are implemented in real, commercially available products.

The development of converged voice and data networks and, most especially, Voice-over-IP technology has happened in slow increments over several years. The concept of IP telephony has been redefined and changed several times over the course of that evolution.

The first development of the Voice-over-IP concept, or VoIP, was probably in the area of Internet telephony. What many of the early developers were trying to accomplish was the establishment of voice and video calls over the public internet. The main motivation for doing this was getting "free" or at least flat-rate long-distance telephone calls. The most well-known application for this early VoIP application was Microsoft's NetMeeting and other similar applications. NetMeeting allowed a user to set up a voice or data connection to another user and provided surprisingly good (assuming your expectations weren't too high) voice quality. The only major thing lacking was a user-friendly addressing scheme. You needed to know the IP address or at least a domain name of the person you were trying to connect to. Not really as easy or ubiquitous as a telephone number.

In the corporate world, network managers were looking at a similar issue. In many cases, a company's locations would be linked by two or more private links, one for the data Wide Area Network and one for point-to-point voice circuits. These enterprises began to look for ways to combine these links together. In the 1980s and early 1990s the data links were relatively slow, up to around 64 Kbps. It made a great deal of sense to put the WAN links as channels on the voice T1 in that type of environment. From that idea the concepts of the alternate voice-data (AVD) T1 and ISDN-PRI multi-service trunking were born.

As the needs of the WAN bandwidth grew, it became less and less practical to use spare voice bandwidth, so the idea was slowly reversed. The next question asked was, "Can we use this new internet telephony idea to put the voice calls onto the WAN, thus using all of the available bandwidth as one big circuit to be used as needed"?

Out of this question came the first business VoIP application, which was the IP-to-T1 gateway, sometimes called a Telephony Server. As telephony servers were used more and more, they found a niche in providing point-to-point links for both voice and data, especially on overseas tie lines, where they proved and still prove extremely cost effective.

In the late 1990s, several different manufacturers began to expand on the telephony server idea to support not only voice trunks but voice terminals as well. The development of these VoIP offers came from several different industries. The PBX Industry developed its products as extensions to their legacy voice offers. The IP switch and router manufacturers developed it as an extension to their existing IP infrastructures. Other manufacturers developed customer offers on customized PCs or servers.

This evolution has divided the VoIP industry into several different philosophical camps.

Some view VoIP as a subset of voice technology, with voice communication as a specialized tool needing specialized skills to develop.

Others view VoIP as just another application that runs on the IP data infrastructure.

Even others view VoIP as a network type all of its own, different from either voice alone or data alone, needing to be developed separately from legacy devices.

This has led to the development of a number of different VoIP architectures.

Each type of VoIP gateway technology has some fundamental strengths and weaknesses. Many of these strengths and weaknesses grow directly out of the strengths and weaknesses of the organizations that have done the development and also in those particular developers' view on the basic nature of VoIP technology.

PBX/KTS-based Architectures

One of the most common types of VoIP architectures has been developed based on traditional business voice systems commonly known as Private Branch Exchanges (PBX) or, in smaller sites, Key Telephone Systems (KTS).

The PBX is a technology that has been steadily evolving over the last forty years or more, the device almost everyone is familiar with in a business telephony environment. PBXs are, in most cases, highly reliable and possess a large embedded set of end-user, networking, and systems administration features. It is this feature richness and reliability that makes the PBX a strong basis for a VoIP network.

The basic architecture used in nearly all PBX-based VoIP systems is to add some form of network interface, typically some form of Ethernet Network Interface Card (NIC). This interface card plays the role in the PBX-based architecture of the IP media gateway. This card and its control software provide the media and signaling translations between the PBX's backplane and the IP network. Once the signals are on the backplane of the PBX they can use the PBX facilities to access outside facilities and adjunct servers.

Once the IP interface is added to the PBX, the control software of the PBX can be modified to provide the media controller functionality for the IP endpoints in the network. This allows the PBX's software functionality to be extended to the IP network (see Figure 5.1).

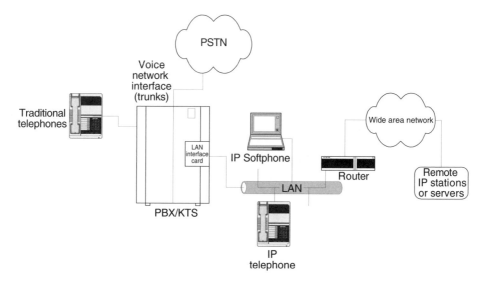

Figure 5.1

Typical PBX/KTS-based configuration.

Strengths of the PBX-Based Architecture

The fundamental strengths of the PBX-based VoIP architecture are essentially the fundamental strengths of the PBX. Most PBXs and their smaller cousins, the KTS, have a highly evolved and robust set of end-user features that allow for highly customizable and flexible telephony environments. PBXs support basic telephony features such as call transfer, call conference, and hold, as well as more advanced features such as call coverage, call forwarding, paging, call bridging, pick-up, and group calling. PBXs, in most cases, also have the ability to support optional, highly advanced features such as automatic call distribution, skills-based routing, and Computer-Telephony Integration (CTI) features.

PBXs have been developed with very sophisticated call routing algorithms, which are used to route calls between multiple locations and also to route calls to and from the public telephone network. Many PBX systems can be used as the basis for complex private networks using ISDN-PRI signaling and even the ISDN-based Q-Signaling (Q-SIG) protocol which can interconnect many different PBXs, and even PBXs from different vendors, into one large "virtual" switch.

PBXs, as they have developed over the years, have become extremely reliable devices. This reliability is enhanced by the multitude of sophisticated management tools that have developed over the years to support PBXs and the vast amount of engineering and technical expertise that exists about PBXs. They have been developed for use by organizations such as airline reservations bureaus, which rely totally on the voice network for their business. Voice systems are designed to meet public safety needs such as support for E911 and other emergency services. In many cases, PBXs share the same architecture and physical structure of the public service telephone switches.

By using the PBX/KTS as the media controller and media gateway on an IP-based network, all of these features are available for use by endpoints on the IP network. This means that features do not have to be re-invented or re-designed for IP telephony. Also, by using the PBX/KTS as the media controller, the user gets a fundamentally reliable, scalable, and manageable server.

Weaknesses of the PBX-Based Architecture

The main weakness of the PBX-based VoIP gateway is that none of these features and none of these tools and resources were built with IP telephony in mind. The problem is that many of these items do not translate well into an IP environment.

For example, many call tracking features including E911 are dependent on the physical port to which a device is attached. In an IP environment, an IP device registers with its media controller using its IP address, not, in most cases, using the address of the physical port to which it is connected. This is done so that the IP endpoint can take advantage of the easy moves and changes offered by switched IP services and such features as dynamic addressing schemes like DHCP. This means that there is often no information in the media controller on the physical location of an IP device and even if there is, the device is often so easy to move that the location information cannot be kept current. This makes the PBX's E911 feature no longer operable as designed for IP stations.

Other examples include inter-PBX networking software such as Q-Signaling (Q-SIG) and ISDN-PRI. Most PBX and KTS systems were designed with hard-wired connections in mind. The control and monitoring of those connections is designed for this type of physical connectivity. Often when these features are extended to an IP environment, the inherent delays involved in IP translation and transit are not accounted for and can cause the management tools for the links to sense a fault and disable the link or otherwise cause the software to behave unpredictably.

Many of these features can be adapted for use in an IP environment, but that adaptation can often negate the advantage of using existing software. The software must often be re-written anyway.

Another problem with using a PBX/KTS-based architecture is that PBXs are closed architectures. In most types they only work with their own interfaces and endpoints and the software is proprietary and cannot be customized easily to suit the user's needs.

The biggest weakness of the PBX-based VoIP gateway is the simple fact that most of the organizations developing and implementing them often have limited expertise in IP networks. Most of these developers know voice networks well enough but don't know the implications of mapping these networks on the IP infrastructure. Also, many of the PBX development organizations do not have the field support and implementation experience needed to deploy voice services on an IP network.

This last weakness is being addressed in many cases by the trend in the telecom industry of the larger telecommunications development companies buying the products, services, and expertise of data networking companies. Some examples of this are Alcatel acquiring Xylan, Nortel and Bay Networks, Lucent and Ascend, and many others.

Migration of Existing PBX/KTS to IP

Another major strength for the PBX-based gateway is that, quite frankly, they are already in place. The vast majority of enterprises have a PBX Key Telephone System (KTS) at most of their locations. The logic here is if you can convert that system into a VoIP controller, you can re-use what you already have and save a great deal of money. This seems simple, but is a bit more complex than it may first seem.

As we discussed previously, the legacy PBX/KTS software was not designed for an IP environment. This means that the software in the PBX must be upgraded to support IP. Fortunately, most PBX/KTS vendors now have IP-compatible versions of their software available.

Assuming there is an IP-compatible version of your PBX/KTS, migration can be as simple as upgrading the software, adding the IP network interface, and adding IP endpoints. This can be a large cost savings, especially in large locations or campuses, because most of the existing voice system can be re-used.

This ability of mixing legacy TDM networks and VoIP networks together is to most users the strongest appeal of PBX-based gateways. This allows users to migrate progressively toward VoIP and deploy new technology where it makes the most sense, like in international voice and data WAN links or remote workers, while still using the existing, proven technology for everyone else.

Having said that, there is still the issue of deploying voice services on an IP network. We will discuss that issue throughout this book.

The migration of an existing PBX/KTS system can often be a relatively simple way to test out VoIP implementations. This allows you to try a few IP connections or endpoints and see how they perform, how they affect your network and what expertise in design and deployment is needed and available to you without making a large investment or committing to new technology all at once.

Router-Based Architecture

Another common type of VoIP Architecture is based on the IP network router. They are a fundamental part of nearly every IP infrastructure. Because they work at the core of the network and are designed to connect multiple protocols and physical interfaces, they are a logical place to develop voice call handling and routing for a VoIP network.

In a typical router-based VoIP network, the traditional IP router is enhanced to act as an IP media gateway. In most cases, that involves adding the transmission and signaling translation applications and support for the standard voice physical interfaces such as ISDN-PRI, ISDN-BRI analog tip/ring (FXS), and analog telephone network (FXO) interfaces to the existing network router.

Once the router is converted to perform the media gateway function it can then be paired with a media controller to provide call control and network services. This media controller application can reside on the router itself or in a separate processor (see Figure 5.2).

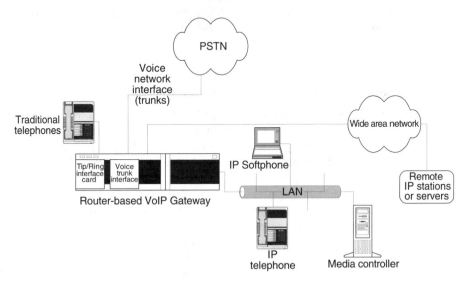

Figure 5.2

Typical router-based VoIP configuration.

Strengths of the Router-Based Architecture

The major strength of the router-based gateway is that is fundamentally an IP-based device. All of the functions of a router are designed to work in a packet-based environment. At the core of this strength is that the developers of these gateways have a vast amount of expertise on the engineering, deployment, and support of IP networks and applications.

Another major strength of the router-based architecture is that it can be a highly modular design. By using published standards such as MGCP or Megaco, the router-based gateway can interwork with a wide variety of media controllers. It is this ability to work through standards that gives the router-based architecture its flexibility.

The potential for modularity and customization has been the biggest selling point for the router-based architecture. The idea is that network managers could pick and choose their gateways, controllers, and adjunct applications and even the terminals from multiple sources and build exactly the type of network with exactly the type of functionality they wanted. This is fundamentally the basis behind the whole concept of open, standards-based architectures.

Probably the biggest strength of the router-based gateway architecture is the large amount of available knowledge and expertise available on IP routers. Through the wide availability of training and certification, especially training involved with Cisco Systems certification programs, there are a vast number of available, qualified IP network engineers.

Weaknesses of the Router-Based Architecture

The main weakness of the router-based architecture is that all of the user features, call routing features, and administrative features must be created for this new media controller application that works with the gateway. This means that anyone who is developing a system using this architecture must re-design even the most basic of telephony features.

In addition to developing this new feature software, the media controller for this type of architecture has to be built in such a way as to be reliable and scalable enough to support large, complex voice and, eventually, video networks.

These last two issues are the reason that most first-generation router-based architectures have been designed around smaller, basic-functionality offers.

The largest weakness of the router-based architecture is the mirror image of the largest weakness in the PBX-based architecture. That is, most of the organizations developing these router-based VoIP offers do not have much experience with the design, development, and deployment of voice services. The same is also true of the field engineering staff, who work with IP routers often and do not have experience with voice and video service issues.

PC-Based Architecture

The third major type of VoIP network architecture seen commonly today is a PC-based architecture. This architecture, as the name implies, involves developing the basic converged IP network architecture on a Personal Computer or server or a group of PCs or servers.

This architecture is built by building media gateway software for the PC and using it to control either off-the-shelf or customized physical interfaces for the links to the various networks or devices. Along with the media gateway software, media controller software is written to provide control of the media gateway and to provide calling and management services. In many PC-based architectures, the media controller and media gateway software are co-resident on the same box, but they can also be on separate PCs (see Figure 5.3).

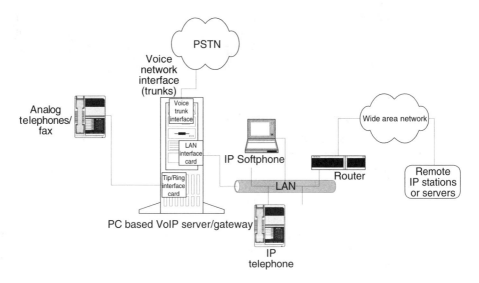

Figure 5.3

Typical PC-based VoIP server.

Strengths of the PC-Based Architecture

The strength of the PC-based architecture comes from its low-cost and highly customizable nature.

The potential low cost of the PC-based VoIP servers comes from the fact that most of the parts can be bought "off the shelf." PCs themselves have become fairly standardized and low cost. There are a large number of available PC adapters for all of the common gateway links. Ethernet NICs, analog station, analog trunk, and digital trunk cards are widely available from many manufacturers.

The customizable nature of this architecture is derived from the fact that these devices are not dependent on any particular type of PBX or router and can be developed independently and work with legacy elements through open standards. Also, because the software is written on an open platform like the PC, it can be easily customized for the individual network and user requirements.

This customizability also gives the PC-based architecture, like the router-based architecture, the ability to interwork with off-the-shelf components such as other media gateways and adjunct servers through open-standard interfaces.

Because of this low cost and flexibility, most of the PC-based architectures available commercially today are aimed at the smaller network market. This market tends to be very price conscious so they are drawn to the inherent potential savings of a converged network, especially when the network elements are relatively low in cost.

Weaknesses of the PC-Based Architecture

The basic weakness of the PC-based architecture is the fact that it is PC-based. PCs are often not as reliable or scalable as dedicated network devices such as PBXs or routers.

The other main weakness is, like the router-based architecture, all of the call handling, user feature, and network management software has to be written from scratch. This is one of the main reasons that the initial generation of PC-based gateways offers limited user features, until those features can be developed further.

Specialized Gateways

The VoIP industry also supports a number of specialized gateway products for use fulfilling specific user applications or network niches. Some of these type are

- Telephony gateways
- Multi-point switched gateways
- Remote telephony servers

Telephony Gateways

"Telephony Gateway" is used here as a generic term for a network device that takes a legacy voice link and converts it to IP for transport across the IP WAN. Most of these devices are designed to work with PBX-to-PBX digital tie trunks so that those trunks can be routed over the data WAN.

Technically this device is a simple media gateway, media controller in a box used for obtaining the cost saving of a converged WAN without changing any of the voice infrastructure at either end. Many variations of this device are available commercially either as stand-alone devices or as part of a router or multiplexing device.

Multi-Point Switched Gateways

The multi-point switched gateway is an interesting twist on the telephony gateway concept. In this case, the gateway converts the voice trunks into IP for transport across the WAN, but this gateway also has a PSTN interface (see Figure 5.4).

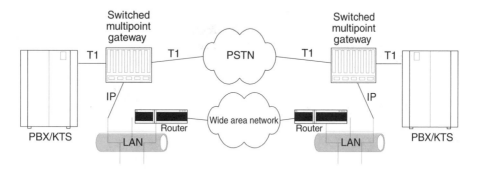

Figure 5.4

The multi-point switched gateway.

What this device does is monitor the performance of the IP WAN. When it sees that the IP WAN is congested enough to degrade voice performance, the gateway re-converts the signal back to digital or analog voice and transmits it over the PSTN. This gives the cost savings of putting the trunks onto the IP WAN, while providing a means to re-route the call to a PSTN service if the quality of the WAN link cannot be maintained.

Remote Telephony Servers

The remote telephony server is another specialized form of media gateway. This device is designed to allow IP-based remote devices to access legacy voice systems.

The way this is designed is that the remote IP device accesses a specialized media gateway which is connected to a voice port on the legacy voice system. This allows the functionality of that voice port to be extended to the IP endpoint without changing the internal structure of the legacy voice network (see Figure 5.5).

What has happened here is that, for all intents, the legacy PBX sees the devices as one of its voice points and that voice points physical interface is simply converted to IP for transport to the physical IP device.

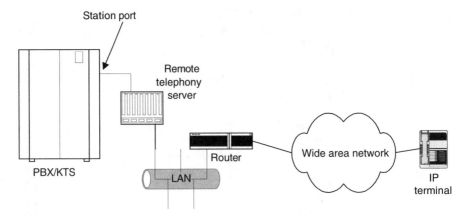

Figure 5.5

Remote telephony server.

Converged IP Network Evolution

While many VoIP manufacturers tout their products by mentioning their potential to work in the open, standards-based environments, the truth is that most of those open standards are not yet finalized well enough to allow for much inter-vendor connectivity. This means that almost all VoIP implementations today, whether they are PBX-based, router-based, or PC-based, are closed, vendor-specific environments. It is true that the IP infrastructure and the VoIP devices can be from multiple vendors. However, the media gateways, media servers, and IP terminals are almost always from a single vendor in any given implementation.

Standards are beginning to stabilize to the point where many of these manufacturers are beginning to modify their products to support industry standards. SIP is starting to look like it will be a standard supported by many manufacturers, and is expected to allow for a much a greater ability for SIP-based terminals to work with multiple types of media controllers. Megaco/H.248 is beginning to take hold in the industry as the protocol that media controllers are using to connect to media gateways, first in a single-vendor environment, but multi-vendor interworking is beginning to show promise.

The development of open standards and developments to overcome the inherent weaknesses of any particular architecture is having an interesting effect on the evolution of these different architectural philosophies.

Manufacturers of PBX-based VoIP offers are beginning to offer dedicated media gateways that connect back to the main PBX processor using MGCP or

Megaco. In parallel with these new offers, the PBX processors themselves are beginning to be modified to work in more open environments. Many PBX processors are also being re-designed to work more closely with IP networks supporting protocols like SIP and H.323 as a core part of their architecture. A large number of PBX manufacturers are even moving to standards-based processors for their main control to better facilitate this more open architecture.

Meanwhile, the manufacturers of router and PC-based VoIP offers are developing more and more user, network, and systems features into their products and beefing up the media control processors to be much more scalable and reliable. At the same time, router-based gateways are becoming more scalable and need more processing power as they are beginning to be moved out of existing router products and into specialized media gateway boxes.

The interesting side of all of this evolution is that as you look at the architectural directions of these different offers, the end results all look remarkably the same. Any reliable scalable, open, media controller, controlling through a standards-based interface through a set of media gateways, services open-standards-based endpoints (see Figure 5.6).

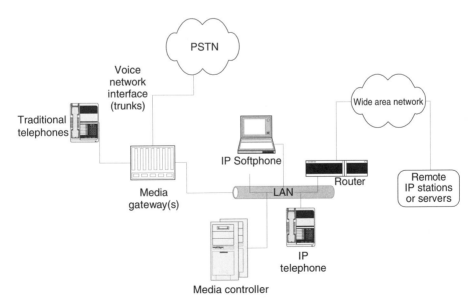

Figure 5.6

Endgame (open serve, with open gateway, standard terminals and interfaces).

All of these diverse philosophical approaches lead to the same conclusion.

In the end, the only difference is if the media gateway looks like a PBX cabinet or an IP router and the only way one manufacturer has to differentiate their product is the applications and capabilities of its software.

In this chapter, we have discussed several common, real-world implementations of a converged VoIP network. Each of these approaches has several important strengths and certain inherent weaknesses. Each is a method commonly used in commercially available products today. As each of these architectures develop into the future it is notable that they are all evolving to more closely resemble the theoretical VoIP architecture discussed in earlier chapters.

CHAPTER 6

VoIP CLIENTS

There is really only one portion of the converged network that the average user sees, feels, and touches. This is the device that determines user acceptance of the new, converged environment. In a Voice over IP environment, that device is the IP voice terminal. For this chapter, we will discuss the many variations of the IP telephone from a PC-based soft telephone to a physical IP desk telephone. In the course of this discussion we will look into the unique features and capabilities of each of these telephone types.

IP voice terminals come in a wide variety of configurations. They range anywhere form a simple H.323 application on a PC to the most elaborate multi-button, multi-function IP desk phone with integrated access to both voice and data applications.

In the traditional telephony architecture, the telephone is hard-wired to the voice server or public network so the only information the telephone can access is what can be directly reached through the voice network. An IP voice terminal, on the other hand, is directly connected to a converged voice and data network and can potentially have access to all of the applications and services available on the IP network.

The other unique characteristic of the IP voice terminal is that it is not hard-wired to the server. The IP voice terminal is connected to the switched IP infrastructure. This gives an IP-based telephone much greater flexibility on how it can be connected across the network. This switched connectivity along with some of the flexible endpoint tools available in an IP network, such as Virtual LANs (VLAN) and dynamic addressing (through applications like DHCP and DNS), give IP voice terminals more of a "virtual" connectivity rather than a fixed one.

The two primary characteristics are the ability to access the IP network applications and "virtual" connectivity. These give IP voice terminals capabilities that are often far beyond the capabilities of traditional telephones. Here are a few examples of the advanced capabilities available on some IP voice terminals.

Remote Access: By registering to the media controller over the IP WAN or through some form of remote access, the remote user can have all of the same capabilities as a locally connected terminal. This enables complete virtualization of the voice network. This application is becoming increasingly popular in the contact center environment as it allows for fully functional remote and home call center agents.

Multi-Service Connectivity: The IP voice terminal device can combine voice connectivity with applications such as web access on a single device to give access to web applications such as corporate directories or web-based messaging services.

Integrated Contact Center Agent: By using a device that provides telephony as well as access to customer records, contact managers, messaging, and other customer service applications, a truly integrated customer service workstation can be deployed to handle a wide variety of customer service tasks.

Integrated Directory: Because the IP terminal can access a directory application on the IP network, users could potentially use the entire on-line corporate directory as a "speed-dial list."

Desktop Integration: IP voice applications can be integrated with other desktop applications such as contact managers, calendars, and workflow managers to add direct telephony connectivity to these applications.

The list of specialized applications that can be enabled through an IP voice terminal is obviously longer than can be given here, and the new list of applications is constantly growing.

Because of this wide variety of potential applications for voice connectivity, IP voice terminals come in a wide variety of forms. These forms can be broken down into one of three major categories—IP Telephones, Soft Telephones (Softphones), and Thin Clients.

Technically, the IP telephone is a thin IP client optimized for telephone call handling. For the purposes of this discussion, we will use the term "IP Telephone" to describe a physical device that acts in the same role as a traditional voice telephone as its main purpose. "Thin Client" in the later section refers to a software-based thin client like a web-browser-based IP telephony application. We will discuss each of these separately.

IP Telephones

The IP telephone comes in a wide array of shapes and sizes. It can look like anything from a traditional single-button or multi-button telephone to a highly elaborate, screen-based device. Some even have touch-sensitive screens or pull-out keyboards to access special features.

In most cases the IP telephone, in the enterprise environment, is designed to look and function very much like a traditional multi-line business telephone. This design usually has been modified to add some of the extra capabilities that an IP phone can access, such as directory access or a web browser.

Because the industry standards for VoIP have not been universally designed and implemented, nearly all IP telephones are currently specific for working with a given media controller. As standards such as Session Initiation Protocol (SIP) are finalized and gain greater acceptance, more and more off-the-shelf IP telephones may become available.

Unlike a traditional digital business telephone, an IP telephone needs to be an intelligent device. The IP telephone must interact with the IP network infrastructure just like any other IP endpoint. This means that the IP telephone must either know its own static IP address or be able to obtain one through regular IP network services such as DHCP and DNS. After the IP telephone establishes itself on the IP network, it must also be intelligent enough to locate and register with its media controller. Once it has registered with its media controller, the IP phone must also have enough on-board intelligence to participate as the terminal device in the appropriate VoIP protocol (H.323, SIP, etc.).

IP telephones have specific needs that are very different from the needs of a traditional telephone. These needs make issues such as connectivity, power, and management somewhat more complex than their legacy voice counterparts.

Connectivity

IP telephone connectivity is very different from traditional telephone connectivity.

In a traditional analog or digital telephone, the telephone device is hard-wired through house cabling all the way back to the voice server. This cabling is separate from the cabling used in the local area network (see Figure 6.1).

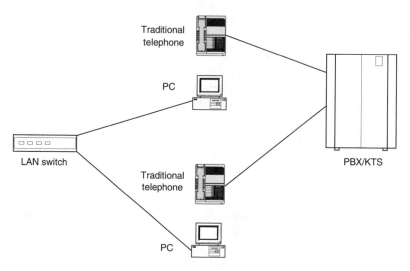

Figure 6.1

Traditional telephone connectivity.

An IP telephone, in contrast, is typically only connected to the local IP switch. That switch is then connected to the other switches and routers in the overall IP network. The IP telephone accesses its media controller through a switched IP connection traversing the local and/or wide area infrastructure. An IP telephone connection in a typical IP local area network is illustrated in Figure 6.2.

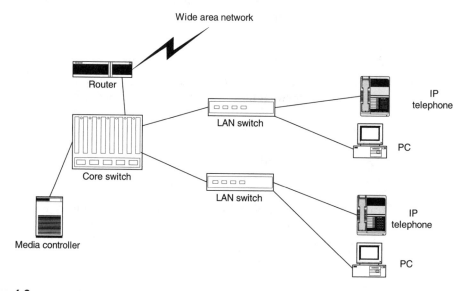

Figure 6.2

IP telephone connectivity.

There are often many cases in an office environment where an IP telephone and an IP data device like a PC may be located on the same desk. Given that both of these devices need the same physical access to the IP network, this would imply that to deploy IP telephones, the IP network interfaces to these work-stations would have to be duplicated. As a result, this would negate one of the major selling points of VoIP, that is, the idea that you would only need a single cabling infrastructure.

To overcome this problem, most currently available IP telephones support some form of switch or hub, built into the physical case of the IP phone. This device is usually a three-port switch or hub with one port connected internally to the IP telephone, one to the IP network switch, and one to the local computer. This ability to provide for single cable access to the desktop gives several options for connectivity to the IP telephone.

The first option is to connect the IP phone and the data device (computer or workstation) to the local switch (see Figure 6.3). This is the most common connectivity option and aids in rapid deployment with minimal modifications to the existing environment. This arrangement has the advantage of using a single port on the switch to provide connectivity to both devices. Also, no changes to the cabling plant are required if the phone is line powered (see the "Power Issues" section). The disadvantage is that if the IP phone goes down, the computer also loses connectivity.

The second option, shown in Figure 6.4, is to connect the IP phone and the computer using different switch ports. Although this option doubles the switch port count for every user, it provides a level of redundancy for the user. If the phone goes down, the PC is not affected and vice versa. Also, you can connect the phone and PC to ports on different switches or switch modules, thus achieving another layer of redundancy by providing protection for one of the devices if either module goes down.

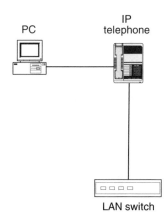

Figure 6.3

Single-port IP telephone/PC-connectivity.

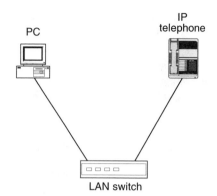

Figure 6.4

Two-port IP telephone/PC connectivity.

Power Issues

Traditional analog and digital phones are almost all powered over the connection to the PBX or central office. This is often referred to as "Phantom Power." The main advantage is that these telephones did not need access to local power or any power backup that supported the PBX or central office (who would also power the telephones).

In an IP network, there is traditionally no provision given to power the end devices. Most IP end devices such as computers, servers, and data terminals are locally powered devices. Also each switch, hub, and router in an IP network is powered separately so there is no central source of power throughout the network. Because of this lack of power, one of the main issues in deploying IP telephones is how to power the phones. There are 3 main options for powering IP telephones.

- Local power
- Powered patch panels
- Inline power (power over Ethernet)

Local Power

Local power is quite simply plugging the telephone into a local power outlet using some form of local power adapter (see Figure 6.5). This has the advantage of not requiring any modifications to the IP infrastructure.

The main disadvantages of local power are that there needs to be a local power source at every telephone and there is no convenient way to provide backup power to all of the individual telephones.

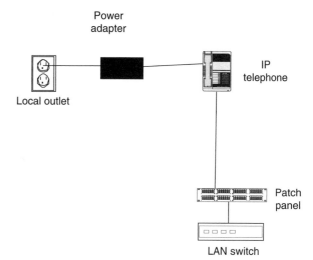

Figure 6.5

Locally powered IP telephone.

Powered Patch Panels

A powered patch panel, sometimes known as a mid-span power supply, is a device that is connected between the local switch and any device that requires power. Typically, the patch panel will provide this power on one or more of the unused pairs on the Ethernet cable (see Figure 6.6).

The major advantage of this power scheme is that the powered panel can be located in a wiring closet so no power source at the desktop is needed and the patch panel can have an associated UPS in case of power failure.

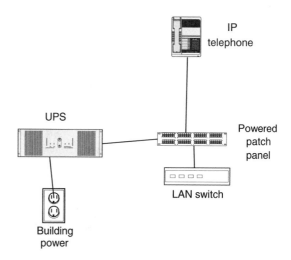

Figure 6.6

IP Telephone power via powered patch panel.

The downside of this power scheme is that the patch panel only connects to devices that need power and care must be used not to add power onto cables connected to devices that may be damaged by the power on the cable, such as a PC NIC card. This adds to the burden of moving and changing both IP telephones and computers on the LAN.

Inline Power

Inline power is a method of applying power directly from the local Ethernet switch. This power is delivered either as a DC voltage on the Ethernet signal pair or on the unused pair of an Ethernet cable (see Figure 6.7). The key aspect of this type of power is that there is some mechanism in place to determine if power is needed by a given device connected to the switch and if it is delivering power only to those devices that need it.

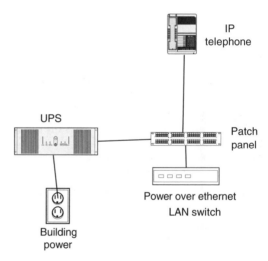

Figure 6.7

IP Telephone power via powered
Ethernet switch.

Again, the advantage of this is the switch can be located in a wiring closet, so no power source at the desktop is needed and the switch can have an associated UPS in case of power failure.

The disadvantage here is that the current power over Ethernet offerings are proprietary, so only certain brands and models of IP telephones can be used with certain power-providing IP switches.

There is a proposed IEEE standard for inline power over Ethernet called 802.3af. We will look into 802.3af in more detail, as most vendors have announced plans to support the 802.1af standard for Power over Ethernet (PoE).

802.3af

The Institute of Electrical and Electronics Engineers (IEEE) formed a task force to standardize an approach for passing power over Ethernet. The task force, called 802.3af, is being run under the auspices of the 802.3 workgroup, the same group that developed the original Ethernet specification.

This standard is expected to be finalized in early 2003. Power over LAN will take the first step toward enjoying the same benefits as any open system. There will be a wide range of vendors providing Power over LAN solutions, hopefully providing customers with multiple supply sources.

The 802.3af is designed to support 10 and 100 Mbits/s Ethernet operating over category five unshielded twisted pair (UTP/FTP) cabling—the kinds of Ethernet commonly used in corporate networks.

The basic network architecture and the underlying cabling of the 802.3af-compliant network remains the same as a standard 802.3 network, yet there are four significant differences.

In order for the network to carry power, one needs to add power sourcing equipment (PSE). This is basically the source of power and the means to integrate that power onto the network. Not only does the PSE inject power into the Ethernet network, it also provides a detection method for determining whether the Ethernet device on the other end of the cable, the Powered Device (PD), is 802.3af compliant or not. The PSE technology can either be implemented outside of the existing switch, a technique called a midspan solution, or inside the switch, called an end-span solution.

Attached to the PSE is the UPS. Today's networks rely on discrete power-backup solutions. A UPS is connected to each device that requires alternative power. With the Power over LAN solution, this function is centralized in a UPS connected to the PSE, which in turn may require further changes in the environmental conditions of the room needing to support this UPS with all of its electrical and cooling requirements.

Management may also be added to monitor and control the PSE. This management function may be integrated into a standard network management platform using the simple network management protocol (SNMP) or through a custom platform. Beyond the basic control of the PSE, the management station provides additional power management functions, such as power quality of service (QoS), where key users are given higher priority to power in the event of an outage.

At the periphery, if required, the device, like an IP phone, can use a splitter to separate power from the Ethernet signal. This splitter can be integrated into the device or for new devices can be situated in the wall or the desktop. No other changes are required to the network. The cabling and all existing Ethernet devices remain unaffected.

Wiring presents the biggest challenge to the 802.3af networks. It's the wiring (and patch panel connectors) that impacts the amount of current that can be carried and how the power can be inserted into the line. Although Ethernet cabling can handle 1 to 2 amps of power, about enough power to drive an inkjet printer, the actual implementation of 802.3af looked at the weakest link, the patch panels and various connectors, and limited the power to 15.4 W (44 V to 57 V) from the PSE. Reducing the power loss on the cables at the 802.3 limit of 100 m, one arrives at a maximum of 12.95 W available for the PD. Many applications are satisfied with this power level. It is even sufficient to charge a laptop battery, while on standby.

The last thing a network manager wants is to start changing the configuration of existing Ethernet devices. Ensuring that devices can remain attached to an 802.3af network without being damaged requires a robust detection mechanism for identifying whether or not the installed device is 802.3af compatible. What's more, the mechanism must identify the devices as 802.3af compatible without modifying the end-node.

The 802.3af calls for resistor detection. The PD incorporates a circuit, which presents a 25 K Ohm resistor to the line. When the power source injects current into the line, the device measures the voltage on the input line. If the resistor is identified, the PSE turns on the 48 V power, and the PD automatically connects the DC/DC stage and powers on. If no resistor is detected then power is not supplied, thus ensuring that power-sensitive devices are not damaged.

The true power of this is that 802.3af-capable switches or patch panels can be distributed across the IP infrastructure assuring the ease of moves and changes that are a large part of the appeal of IP-based networks.

Moves, Adds, Changes

A major selling point of an IP telephone is the ease with which IP telephones can be deployed and moved from one location to another. This feature is directly derived from the fact that the IP telephone is not directly hard-wired to a physical port, but is connected "virtually" through a switched IP network.

In a traditional voice network a telephone's identity, its extension number, and all of the features and functions programmed for that specific phone are associated directly with the physical port on the server to which that particular phone is attached. What this means is that if a systems administrator needs to move that phone to a different location that phone need to be re-associated with the server in one of two ways.

First the cabling that is connected to that phone's particular port can be redirected through patch panels or punch-downs to the new location of the telephone. This obviously requires some physical intervention by the administrator. The second choice is to change the voice server software to move the translations re-associating that phone's identity to a port that connects to the new location. Again this requires manual intervention.

Many PBX vendors are trying to overcome this obstacle by adding special codes that can disassociate the phone from one port and automatically re-associate it with another port.

In contrast, an IP telephone's identity is typically stored in memory or firmware on the phone. When the phone registers with the media controller, the media controller can then map that phone's feature set and address (extension number) to the telephone. This mapping, in most cases, is done via the IP telephone's IP address. Because the association of the phone to the media controller is done at an address basis rather than a physical basis, moves and changes are greatly simplified. Much like what can be done with a laptop PC, an IP telephone can be moved by simply disconnecting the telephone from the IP switch port, moving the phone, and plugging it into another switch port. Once the device is re-attached, it can re-register with the media controller and the phone has all of its previous functions moved with it. This phone has now been moved without systems administrator involvement.

This freedom of movement has allowed some enterprises to deploy telephones in new ways. A telephone can be associated with a person and that person can plug the phone in wherever they need to and get access to their extension and calling feature. This capability is often used with virtual or softphones, which we will discuss in depth later.

This freedom of movement can cause problems within an enterprise in that there is no control of the movement of telephone devices. Ironically, many administrators are now starting to deploy tools such as switch port restrictions and VLANs to limit the movement of these telephones. The goal of these devices is to block the ability of the IP telephone to register with the media controller after it is moved, hopefully discouraging users from moving the phones.

Remote Support

Another unique attribute given to IP phones by virtue of their "virtual" connectivity is the ability to deploy IP telephones remotely from the media controller. As we discussed before, an IP phone registers to its media controller over the switched IP network and then all of its extended features and attributes are assigned to the telephone. If a given IP phone can access the media controller over a wide area network or over some form of remote access, then that phone could register on the VoIP network the same as if it were local to the media controller.

The real key to the popularity of this feature is that it is possible for that remote telephone to have all of the same feature functionality and access as any locally connected telephone. This opens up new applications and the concept of a virtual voice network. Some of the leading applications that are often used to exploit this functionality are

- *Remote/home office workers*—A worker at a remote or home office can connect via remote access back to the main IP network, register with the media controller, and extend their telephone with full functionality to the remote site.
- *Remote contact center agents*—Because a remote access IP telephone can, in theory, have all of the same functions that a local telephone can have, the functions of a contact center agent can be extended to a remote site as well. This has great applications in the use of temporary and part-time agents, saving on physical office space and access to higher-skilled and/or lower-cost agents in remote locations.
- *Small/Branch office support*—Another application opened up by using IP telephones is the ability to support a small remote location off of a larger site simply by deploying IP phones that register back to the main site. This provides for telephony needs at the branch office without having to deploy any remote telephony hardware or PSTN access.
- *Traveling workers ("Road warriors")*—Many enterprises have many personnel who travel frequently, but need to have full access to telephony features. By using some form of remote access (dialup, VPN, etc.) and an IP telephone, the user can register to the media controller over the remote IP access and get access to all of the telephony features. This essentially allows them to take their office telephone capabilities with them wherever they go. This is a common application for IP soft telephones, which we will discuss in a later section.

E911 Issues

In the last two sections we discussed some of the benefits derived from the virtual connectivity used by IP telephones, mainly ease of movement and robust remote support. There are a few problems caused by this as well. The most commonly discussed and potentially most serious of these issues is the support of network-based emergency calling services, known in the United States as Enhanced 911 or E911 service.

E911 service works by using a database that maps the calling telephone number of a 911 caller with the physical location associated with that telephone. With a home or public telephone that database is maintained by the local carrier and is simply a map of telephone numbers and their physical locations. When a caller dials 911, the call is directed to the PSAP and the location data associated with the caller (often called an Automatic Location Identifier or ALI) is sent to the PSAP agent to be distributed as needed to assist in emergency response.

In a commercial business environment, this works somewhat differently. Most businesses have some sort of PBX or KTS that is directly connected to the public network. The ALI is usually only associated with the business location, not the location of any given extension inside the business office or campus. There are many tools available to the enterprise business to overcome this. Most of these involve a local database that maps internal locations to extensions and then passes this information on to an internal emergency response location or on to the PSAP via ISDN-PRI trunks with station ID, or via specialized E911 trunks called CAMA trunks. Like the public version of E911, these databases map the location of a caller based on the physical location of the telephone.

As we discussed previously, an IP telephone is not associated to the server by a physical location but by a virtual address. This makes it easy to re-locate IP telephones, but it also makes it nearly impossible to accurately maintain a record of the physical location of the IP telephone. You can imagine the consequences of a caller dialing E911 from a recently relocated IP telephone and the emergency response being directed to the previous location of the telephone.

This one feature issue with IP telephones is viewed in the industry as one of the major factors slowing the deployment of IP telephony, especially in larger corporate environments or in government and education agencies. The risk to public safety and the associated liability risk are often a major reason why organizations like that are slow to deploy on-campus IP telephony. There have been a number of proposed solutions to this problem.

One method is to add a mechanism to the registration of the IP telephone to the media controller that updates the MAC address of the IP switch port to which the IP telephone is connected. That address could then be mapped to a physical location and the phone could then be pinpointed to within a 100 meter area (the maximum length of a Ethernet cable from switch to terminal). There are still issues with coming up with a scalable mechanism for supporting this, and this does not address situations like remote access servers and wireless IP.

Other more exotic methods of tracking phones have been proposed, up to and including adding a locator device such as Global Positioning System (GPS) chips to IP phones, but so far none of these has proven practical. The only current solution is to use devices like port restrictions to control the movement of IP phones (but that defeats some of the benefits of IP phones).

Overcoming this issue is still one of the biggest impediments to more general acceptance of IP telephony.

Softphones

As we have just discussed, one of the core attributes of the IP telephone is the virtual-ized association (as opposed to the hard-wired connection of a traditional phone) that it has with the media controller. The next logical step along that path is to make the IP telephone itself a virtual device on the network, that is, making it a software element that can be part of another device, service, or application. This type of virtual telephone is known by many names. It is sometimes called a soft telephone or virtual telephone. In an MS Windows environment it is often called a Windows Telephony (WinTel) application. For our purposes we will call it by the generic term IP softphone.

The typical softphone consists of four major parts: some form of host device like a PC, laptop, or palmtop computer; software to support the VoIP protocol; the user interface software; and a device like a PC soundcard and microphone or the equivalent to provide physical input and output of the user's voice.

Platform Considerations

One of the first things that needs to be considered when thinking about softphones is the platform on which the software is supported. The vast majority of softphone applications available in the market today are MS Windows-based applications. These applications are designed to run on a standard business personal computer or laptop. There are a growing number of softphone offers that are starting to come to market on Windows CE (Pocket PC) and Palm0S-based devices as well.

For all of these types of platforms there are some common capabilities that are required to support a softphone.

Computing Power—Softphone applications require the PC to handle both the protocol handling like H.3232 or SIP as well as the user interface. This requires a reasonably robust computer with sufficient memory to support the application. Typically a modern, business-class PC or laptop is sufficient.

Sound Processing—Natural conversations are full-duplex. That is when both parties are often speaking and listening (or at least trying to) at the same time. This means that one of the requirements of a platform for a softphone is to support full-duplex sound. If a PC or laptop has only a half-duplex sound system, the user can experience clipping and talk-over on calls (similar to the performance of low-cost speakerphones). This should be taken into consideration before deploying softphones as full-duplex. Also, sound cards have only recently started to be deployed in laptops and almost no Windows CE devices have full-duplex sound capabilities.

Speakers/Microphones—A key determining factor in the performance and ultimate user acceptance of softphones is the quality of microphones and speakers available on the user's PC or laptop. Laptop computers especially often do not have terribly high-quality speakers in them. Most softphone applications are best served with some form of PC headset or attached handset to overcome these limitations. There are a large number of available headsets, handsets, and even speakerphones that support PC sound-port interfaces or serial or USB interfaces.

Software Integration

One of the inherent qualities of an IP softphone is that it often resides on the user's business PC or laptop. On this same device, the user may be running their standard business applications such as calendars, contact lists, and customer information databases. This allows the integration of the telephony functions of the softphone with the business applications of the PC.

One of the most common methods of performing this integration is through the Telephony Applications Programming Interface (TAPI). TAPI is a set of standardized instructions that allows access between MS Windows-based applications and telephony features such as calling number ID and telephone dialing. This gives users access to features such as calling directly from a contact list on the PC or bringing up customer information automatically based on the calling-party-ID of an incoming call. These features can also be developed through any number of proprietary DDE applications.

This functionality is often used in contact centers on a larger scale, using calling party information to retrieve caller information from corporate databases and delivering to an agent along with the incoming call.

Another common point of integration is between the softphone and corporate directory applications. This is often done using the Lightweight Directory Access Protocol (LDAP). This can allow the softphone to access a corporate directory and use it for dialing.

Separation of Signaling from Voice

A few vendors in the market are offering a unique application for their softphones. These applications involve using the software to control the functions of a separate voice device like a traditional analog telephone. This is called by many names, including external call control, separation of signaling from the voice bearer, or a "dual-connect" architecture.

This works by the softphone application registering with the media controller as in a normal softphone. What is different here is that the registration contains an instruction to re-direct the voice signal to a separate phone line to provide a talk path. What this then allows is that the functionality of the phone can be the same as any IP-based telephone, but the voice connection does not go through the IP network, thus avoiding the potential quality issues. (See Figure 6.8 for an illustration of this type of connectivity.) In this case, the IP softphone takes control of any line available, like an analog line or cell phone, and uses that as a "long handset cord."

This form of connectivity is especially useful in situations where a user might need the advanced features but does not have access to sufficient bandwidth to support VoIP or there is no reliable QoS mechanism to support them. This is commonly seen in applications such as traveling executives who only have dial

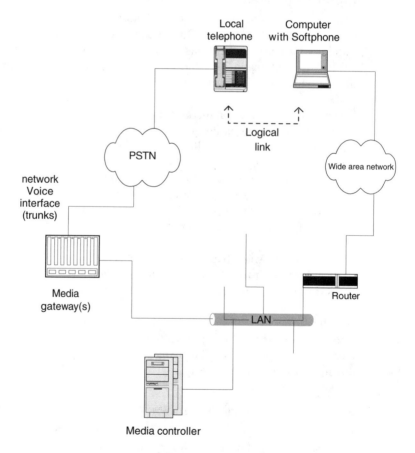

Figure 6.8

IP softphone using voice channel separated from signaling.

access but want full telephone functionality, or home call center agents or service providers that may not be on your own network.

Common Softphone Applications

IP softphones have found for themselves several applications niches that seem perfectly suited for these applications, including the following.

Traveling Workers (Road warriors)—Traveling executives often need access to the advanced features of the in-office telephony network while on the road. A softphone allows them the ability to register with the voice network over the same connection used for other remote access applications like email to establish this kind of telephony access.

Home/Remote Workers—An IP softphone can allow users with remote access to the IP network to also access IP telephony functions from a remote location without the added expense of high-end telephony devices at the remote site.

Call center agents—One common goal in a call center environment is to eliminate as many devices from an agent's workspace as possible. Using a softphone allows the telephony control to be integrated with the agent's computing platform, allowing the elimination of the telephone device from the workspace, but also facilitating integration on the telephony with the call center applications. This could then be applied both to local as well as remote agents.

Wireless Device Integration—A growing number of businesses use wireless devices for in-office mobile staff and for functions like inventory control. Using an IP softphone designed for these devices (for example, a Windows CE device) would allow those devices and their users access to telephony applications.

There are many more applications available and the number is growing rapidly.

Thin Clients

A third major type of IP terminal device is the Thin Client. Technically speaking, all physical IP telephones are a form of Thin Client. That is, they are specialized devices designed especially for a single purpose. For our purposes here, what we are referring to are software-based thin clients, or more precisely, web-browser-based IP telephones.

Basically, this type of IP telephone is really a specialized form of the IP softphone. The major difference is that the application on the client's PC or laptop is a standard HTML browser instead of a custom telephony application.

These types of devices are implemented using a server that works as a proxy to the IP telephony application, a web/telephone server. This server is what registers with the media controller and terminates the VoIP protocols. This server in turn converts this into a session with the client's web browser. See Figure 6.9 for an illustration of this type of device.

The thin client can, in theory, perform all of the functions of the IP softphone and supports the same common applications. The main difference here is that it is essentially independent of the client's platform. The same web/telephone server could support devices running MS Windows, Windows CE, PalmOS, MacOS, LINUX, UNIX, or whatever, beacuse the client is simply an HTML browser. The client would still need some form of support for full-duplex sound

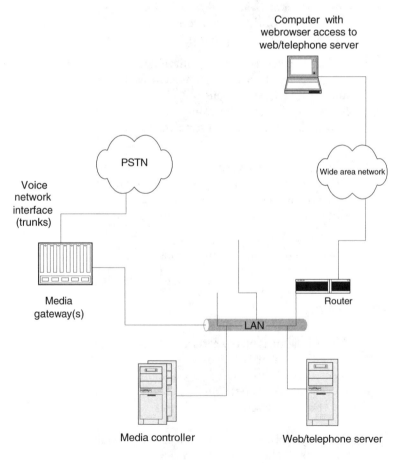

Figure 6.9

Thin client IP telephony.

and some form of microphone/speaker or headset as described in the softphone discussion.

There are also client variations on the "dual-connection" architecture discussed in the softphone section. That is when the web/telephone server would provide a control link over the browser and connect the call over a separate line. This type of arrangement can have many advantages over a traditional softphone, but it does also have some significant disadvantages as well. This architecture is highly dependent on the reliability and scalability of the web/telephone server. If the server goes down it can take many users with it. Just like with any VoIP application, care must be taken to ensure QoS, but in the case of this type of Thin Client the VoIP traffic is HTML traffic and it can be difficult to prioritize the voice application over other HTML applications.

IP telephones come in many forms. Each of these forms—IP telephone, softphone or thin client—has strengths in certain types of applications. In nearly all production implantations of IP telephony most, if not all, of these types of devices are deployed in one form or another. This flexibility is one of the key attributes that will keep driving the implementation of IP telephony on into the future.

CHAPTER 7

VIDEO ON THE CONVERGED NETWORK

Streaming video and video conferencing have been around for several years now. These technologies have often been deployed on dedicated networks. Today, as converged voice/data networks begin to deploy, many users are taking the next, logical step of adding video support to the converged network.

In this chapter we will discuss some of the implications of adding video to the converged network. Some of these implications come from the unique need of video transmissions. Video transmission needs such as performance, bandwidth, and QoS will be addressed.

We will begin by discussing the components and technologies in both video-conferencing and streaming video applications as well as the end-user video devices. After that, we will discuss the need to support these technologies in both a wired converged network and then on the new, emerging wireless technologies.

Video Traffic Characteristics

Video Conferencing

Videoconferencing is one of the simplest yet most popular uses of video technology. For meetings that regularly take place and require face-to-face communication, videoconferencing can substitute for the actual physical presence of remote participants. This reduces travel costs as well as travel time and makes meeting attendance more convenient. It also makes meetings more likely to occur. Frequent and/or ad hoc meetings that might not have been scheduled due to travel costs and timing can be enabled via videoconferencing and enhance the sense of teamwork among people at different locations but working on the same project. Videoconferencing provides remote participants with much of the face-to-face familiarity that comes with physical presence, including elements of facial expression, body language, and eye contact. If videoconferencing is readily available on individual desktops, the cohesive effects of this enhanced communication can be even greater. Collaborative work can then be enhanced further through the integration of videoconferencing with collaborative electronic tools (data transfer, shared whiteboards, shared applications).

It is important to review where we are today with video applications. Understanding the heritage of digital video helps explain some of the limitations in place. The most promising video technology was ISDN. This first widely available digital telephony transport was to be the basis for multimedia calling. Today, this technology is the foundation for most users' current video experiences.

At a high level, ISDN offers the ability to get approximately 100 Kbps, 300 Kbps, or 1 Mb digital streams point-to-point in a high-quality fashion. This is important because videoconferencing takes a lot of bandwidth. Uncompressed analog video is fairly bulky to transport. It can be sent in a simple box with a movie filmstrip in it, but it actually takes on the order of 30 Mbps to deliver electronically. This is roughly what you get to your TV set.

Video Compression

Digital encoding permits the signal to be reduced in various ways. In the early 1990s, it was determined that a digital signal of about 300 Kbps would deliver adequate video for videoconferencing. This is the data rate for videos found in typical room systems today.

On typical videoconferencing systems, the user will not see full TV-quality motion. The U.S. standard for full motion is 30 frames per second. The human eye perceives 30 images per second as smooth motion. This is essential for watching sports, reading sign language, or seeing detailed motion. A videoconference, in theory, doesn't need this quality. A "talking heads" conference doesn't have motion that you would find, for example, in a theatrical movie. In your traditional business meeting, it is rare to see much movement. Instead, we see people sitting in chairs, moving their lips, blinking their eyes, and sometimes making hand gestures. For this, 10–15 frames per second can suffice, with motion traces usually unobjectionable.

These assumptions about how you would use the technology led to the designs you see today: videoconferences with less quality than what you see in your living room, on even an inexpensive TV set. Figure 7.1 shows a typical videoconferencing deployment.

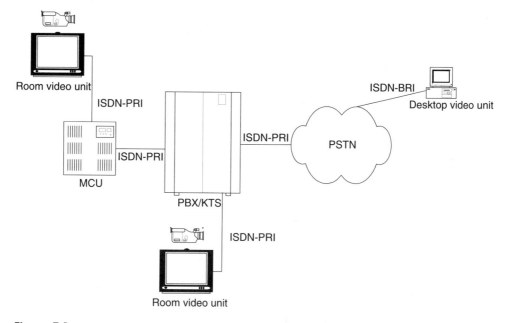

Figure 7.1

Standard H.320 video implementation.

Videoconferencing Standards

ITU H.320 is the standard that enables videoconferencing. It is a sophisticated protocol that permits all of the devices in a video session to manage the session and what each user sees. H.320 includes standards for how a videoconference server will interoperate with video clients. It was written by the ITU, an organization that was set up to define worldwide telephony standards.

In the early 1990s, as it became apparent that video would be carried over packet networks, the ITU began evaluating IP video standards. The team that developed the H.320 definition defined a packet version of this standard, reusing as much of the H.320 standard as possible. This standard is known as the H.323 IP video standard.

MCU

One of the common uses for videoconferencing is to connect together multiple locations, creating a "virtual meeting room" that exists for that particular time and group configuration facilitated by the network. Such "meeting rooms" are created through the use of a Multipoint Conferencing Unit (MCU). The purpose of an MCU is to connect three or more videoconferencing systems in the same conference, managing audio and video from each participant to the others such that group communication is achieved. Data sharing is also possible between all participants in a multipoint conference though current implementations vary greatly in terms of how this is done and also how well it works.

Videoconferencing standards outline two component processes that form the basis of any multipoint interaction—the MC (multipoint controller) and the MP (multipoint processor). The MP is optional and, if present, be more than one may be used. The standard provides users with two different ways to provide multipoint functionality overall: centralized versus decentralized.

The MC provides for overall control of the conference. This involves forming connections between all endpoints, negotiating common capabilities, and communicating to the MP regarding any necessary switching of audio/video sources. The MP handles the actual processing of incoming and outgoing audio/video streams. Audio from all sites in a multipoint conference is typically mixed and delivered back to all sites in full-duplex mode. Video, on the other hand, may be handled in a few different ways, as follows:

1. Switched based on voice activation (everyone sees the current speaker)
2. Switched via manual control ("chair control," where the designated chair decides whose video is being seen)
3. Displayed together on a split screen display (continuous presence)
4. Displayed in individual video windows, one for each site that is being received.

In a centralized MCU, the MC and MP are included in a single unit to which all endpoints connect. This forms a physical and logical star configuration with the MCU at the center. Each endpoint is, in effect, in a point-to-point call with the MCU.

In a decentralized MCU, the component processes (MC and MP) are present to some degree in the client endpoints. The MC of one endpoint will most likely be used to control the conference while each endpoint uses its own MP to send/receive streams in accordance with its own capabilities. The video/audio/data streams from each endpoint are sent one-to-many, which requires the use of IP multicast to facilitate group identification and participation.

Arguments for and against centralized versus decentralized multipoint conferencing are not unlike those surrounding the debate of centralized server-based computing versus peer-to-peer computing. Centralized MCUs are more thoroughly defined and more readily understood; therefore they are more widely available in standardized product implementations. There are inherent pros and cons of each approach.

- Centralized functionality lends itself to improved reliability, control, and management. It also allows for advanced capabilities to be introduced into one entity but made available to all, thereby reducing costs at the endpoints. Of course, cost is then shifted to the central unit, in this case, the MCU. Other functionality, such as additional transcoding or network gateways, can also be fairly readily added to a centralized MCU, extending the service capabilities further than "simple" multipoint call handling. Again, this increases the cost and complexity of the MCU while decreasing cost and complexity required for client endpoints. Another consideration is that, until quite recently, most centralized MCUs forced each conference participant to the lowest common denominator for call capabilities. For instance, if one participating endpoint could only send/receive QCIF calls at 128 K bandwidth, all other participants in the same conference would be forced to send/receive the same. This limitation is changing as increased transcoding capabilities are being introduced into some centralized MCUs.

- Decentralized functionality more readily supports flexibility for end users and a more distributed load over the network. Cost can be determined and distributed based on capabilities desired for particular endpoints. Each endpoint also determines its own send/receive capabilities and does not need to adjust these based on what other participants can do. Also, in addition to providing a mechanism for group calling, support for IP multicast allows the most efficient use of bandwidth as determined by the placement and concentration of participating endpoints within the network.

Another consideration for the implementation of an MCU is hardware- versus software-based. Hardware implementations tend to be more expensive and are likely to contain a variety of proprietary components but are likely to be faster and are also prone to be more reliable. Software implementations are more portable, more flexible, and less expensive but may suffer performance issues due to their reliance on the operating system and resources of the computer on which they are running. Each type of implementation is available on the market today in a variety of forms. A careful matching of performance requirements to cost variables should be combined with a broad comparison of available products within each implementation type before a final buying decision is made.

There are a few different hardware-based MCU configurations that are available today. One type features a modular chassis that holds one or more power supplies and a number of other interface cards. Connection "ports" are included on some of these interface cards, with the number of ports available corresponding to the number of sites that can be participating in conferences at the same time. Other hardware-based MCUs are based on more streamlined units that do not feature pluggable modules but instead are ordered with the desired number/type of ports built in. In either case, multipoint conferences involve specific numbers of endpoints established on the MCU and use as many actual ports as necessary for the number/type of connections and the amount of time required. Some MCUs include scheduling capabilities that allow conferences to be configured/scheduled in advance and brought up automatically. Others only allow ad hoc use of available ports on a "first come, first served" basis.

Software MCUs operate in much the same way as hardware-based MCUs but consist only of a software package running on a powerful server/computer. Software MCU manufacturers usually limit the number of simultaneous connections by a license key that is purchased by the customer. However, there are technical limits to the number of sites that can be connected together at one time based on the processing power and speed of the server.

Both hardware- and software-based MCUs can be connected to allow larger numbers of sites to be conferenced together simultaneously. This is termed "cascading" and is a functionality that is described in the H.323 standard. MCUs from different vendors should therefore be able to be cascaded together quite readily. In order to do this, one of the ports on each of the MCUs is used to connect the other. Audio and video mixing/switching should still operate as if there is only one MCU involved; the cascading is transparent to the participants.

Some concepts that need to be considered in a video conferencing deployment are

1. *Environment of the participants*—Potential environments include desktop, conference room, or auditorium. Will conferees be meeting from their desktops or will a few, small conference rooms be combined into a larger, "virtual" conference room?
2. *Role of voice/video/data*—Is voice transmission sufficient? If video is required, must participants be able to see detail or is video used only to maintain presence? Will graphical materials be exchanged and must participants be able to collaborate using the same application?
3. *Degree of interactivity*—How will the meeting be held? Will there be a single main speaker, as in a classroom environment, or will a productive meeting require that the participants be able to speak to other conferees freely and spontaneously?
4. *More than two participants*—Conferencing between greater than two participant sites will require use of a multipoint control unit (MCU) or multicast network services. In addition to the technical factors, there are also human factors that change the way a videoconference works when it involves more than two parties.
5. *Need for Gateways*—Is there a need to talk with people using endpoints on different protocols? For example, is one endpoint coming in via H.323 (TCP/IP) and another on H.320 (ISDN)? If so, conversion from H.323 to H.320 (and vice versa) will require a gateway to translate the signals.

6. *Video Conference Management*—This not only involves managing the video equipment and network, but also managing and scheduling the conference rooms, MCU, and end-user equipment availability.

Streaming Video

Videoconferencing is a specialized form of streaming video used to support meetings between two or more parties. In this section we will discuss streaming video in a more generalized environment.

Streaming video is used in the distribution of video media across a network. This is often done to send out video data to users on an on-demand basis for training, information distribution, or entertainment. Streaming the video signal, as opposed to file transfer, allows users to begin viewing the material before the entire file is downloaded, thus saving the user from waiting for the entire file download to finish. Streaming video applications often have larger bandwidth needs than in typical business videoconferencing applications.

One of the key aspects of streaming media is the encoding scheme. This scheme determines the quality of the transmission and also the bandwidth needs to support the video transmission across the network.

MPEG

MPEG stands for the Motion Picture Engineering Group. It is an industry-based standards-setting organization that specializes in audio and video compression and transmissions. They currently have three published standards that relate to video compression: MPEG1, MPEG2, and MPEG4. The newest, MPEG4 is on version 2, but its development is continuing. The two other standards that MPEG is working on—MPEG7 (for describing multimedia content) and MPEG21 (which is to define a multimedia framework)—don't currently have a significant role in the videoconferencing world.

The following table summarizes the various formats and their capabilities:

	Typical Image Size	Typical Bandwidth	Max Bandwidth
MPEG1	352 × 240 (std profile)	1.5 Mb/s	2.5 Mb/s
MPEG2	720 × 480 (main profile@main level)	5 Mb/s	15 Mb/s
MPEG4	720 × 480 (main profile, L2)	2 Mb/s	4 Mb/s

MPEG1 is an older standard that was originally designed to compress about 30 m of audio and video onto a CD. It is fairly fast to compress and decompress, but for the number of bits, it doesn't make for impressive video quality. Typically the bit rate is around 1 to 1.5 Mb/s. Because H.263 compression as used in most H.323-based systems produces better picture quality with about the same processing and fewer bits transmitted, MPEG1 isn't usually a contender for videoconferencing systems.

The MPEG2 standard is a commonly used compression scheme for video. There are a number of products that use this format to produce reasonable quality images. While it was developed for broadcast applications, vendors have found that the latencies introduced by using standard definition video, fast encoding and decoding cards allow it to be used in an interactive environment. MPEG2 is a very complex standard that includes many variations on resolution and format that extend from standard-definition TV to high-definition TV specifications. It is the encoding currently used in many consumer TV products such as DVD players, satellite TV receivers, and digital cable TV receivers.

MPEG4 is a newer standard that includes a video encoder that is more modern than the one used in MPEG2. Like MPEG2, it has a wide selection of profiles that range from very low bit rates meant for wireless transmission to very high rates for video editing and exchange. While this standard also includes many other non-video multimedia components, the compression technology is just now starting to appear in video-based products. For lower bit rates and resolutions it can probably easily replace MPEG2 and for the same number of bits provide a better image or use fewer bits for the same quality of image. This is not happening quickly in the TV/broadcast arena because there are already so many set-top boxes and DVD players in people's homes. MPEG4 requires more processing power than either MPEG2 or MPEG1 to encode and decode, which means that in the short term it may not be feasible to use it for real-time interactive work. As with the other MPEG compression schemes, it was developed for broadcast or streaming applications where the latency isn't as much of an issue as it is with videoconferencing. Its application to videoconferencing may only be a matter of time as machines get faster and hardware is specialized to use this standard. The comprehensiveness of the standard, the inclusion of non-video components, and its flexibility for graceful degradation as network conditions deteriorate make it a promising candidate for future videoconferencing applications.

Motion JPEG (MJPEG)

MJPEG, for "motion-JPEG," is an encoding technique that simply jpeg-compresses each video frame before transmission. There are several advantages to this approach, and some disadvantages. The biggest drawbacks are that there are very few products that support it, and it has been seen as a failed technology of the past. Both of these are changing again, and MJPEG is making a strong comeback.

The main advantages of MJPEG are that JPEG-compression is very cheap to do in hardware, is extremely fast, quite robust, and it supports almost any size video frame you want to transmit (subject only to an 8×8 tiling restriction, so images may need to be padded in software). This means you can use MJPEG for video frames from sub-QCIF size up to full-size 1080i-HDTV (1920×1080) and beyond.

Because the codecs are so fast and there is no inter-frame compression used, MJPEG has the lowest latency of any codec currently on the market. This low latency is great for more natural interaction than say H.261, H.263 (the most common H.323 codecs), or MPEG. In general, MJPEG systems also perform much better for lip-synchronization than most H.323 or other codecs. Why this should be so is unclear, but the lower codec latencies do allow for greater flexibility in the decoder and playback setup.

Because MJPEG uses the normal JPEG tiles, and no inter-frame information, errors or packet loss on the network will only impact a tile, or row of tiles, in an image, and the error does not propagate for several frames. This makes it a very sensitive network-testing tool!

A common approach with MJPEG codecs is to give users two choices of compression. One sets a fixed compression level, or "Q-value" in JPEG terminology, for the DCT codec and just compresses each video frame and transmits it. While this provides a very stable and high-quality video stream, it can use up bandwidth very quickly, especially if there is a lot of fine detail ("high frequency data") in the images. An alternative approach allows one to set a bandwidth cap, and the Q-value is allowed to vary frame by frame to meet the bandwidth cap. This means the image quality will vary somewhat from frame to frame. For normal videoconferencing situations where there is little change, though, this is not perceptible.

Because MJPEG compresses on a frame-by-frame basis, it automatically picks up the frame rate of the raw video so there is no issue with handling NTSC, PAL, SECAM, or any other codec, all in the same system. Conversely, some products allow you to specify a particular frame rate (as a fraction of the source stream) and again use the Q-factor versus bandwidth arrangement to fine-tune the performance of a video link.

Another advantage for MJPEG-encoded material is that it is always frame-accurate, which makes editing a lot easier—unlike inter-frame compressed material, where you need to work from keyframes, or decompress and recompress intermediate frames with attendant loss of quality. A variety of software video-editors use MJPEG for on-disk storage for this reason, only converting to MPEG or other format when writing to tape or VCD/DVD.

A more recent change is the support for the newer JPEG-2000 standard. This has moved away from the DCT compression used in JPEG, and also H.261/H.263, MPEG and others, toward "wavelet" compression of image data. The compression achieved is significantly better. It has more flexibility, better error resilience (one of its applications is in wireless video transmission to low-bandwidth devices), and you can embed metadata more easily with the image data. At this stage, support for MJPEG2000 is still in its infancy, and all done in software, but it has some significant potential.

All of these standards can be used to stream video for business or entertainment purposes. Supporting issues for streaming video in the converged network are similar to those for supporting videoconferencing, except that streaming video is less typically less sensitive to delays.

Video Endpoints

As with IP telephones, one of the most important factors in the users' interaction with a video network is the endpoints that they come in actual contact with. There are several variants available, ranging from simple desktop PCs with video cameras to elaborate conference room systems with multiple screens, multiple cameras, and various other media inputs.

Room Systems

A room system is often a vendor's top-of-the-line product designed to provide medium and large corporations, government, and educational institutions with custom room configurations. These systems are used in situations requiring high-quality video performance and extensive conferencing capabilities for applications such as distance learning, boardroom conferencing, and high-performance multi-site conferencing. Performance is based on providing IP data rates as fast as 2 Megabits per second. These systems typically support multiple video and audio inputs such as document cameras, computer video converters, and VCRs. Support for one to three video displays is also common. Complete systems may come with two or more 32" viewing monitors or can be used with larger projection systems. These systems can be permanently built into a room or set up as semi-mobile units that can be taken to another room if necessary. Some units even have built-in streaming services. The main advantages to these systems are video quality and built-in multipoint services. The main disadvantage is price; these systems are typically very expensive.

Set-Top Systems

The term set-top system usually refers to a vendor's mid-level product designed for small to medium-sized conference rooms. This system can look just like the room system with monitors and carts, or it can be a smaller unit that sits on top of a monitor. These systems typically will perform the same as the room systems at lower data rates; however, the set-top systems typically do not support data rates above 768 K and do not have the built-in multipoint conference capability. One advantage to the set-top solution is that products on the lower end of this pricing scale can give a user excellent conference room or classroom performance as long as they do not require the extra services or options offered by the room systems. The disadvantages are that multipoint services would need to be provided elsewhere.

Desktop Systems

A desktop system, simply put, is a videoconferencing terminal, either hardware or software, that interacts directly with the personal computer on a desk. It is designed to be a one-person unit, making it unsuitable for a conference room environment; but it can be very useful for an office or classroom/lab-station solution. It will either connect through the USB port or will have a hardware card to be installed. The speed of the computer processor will directly affect performance. The advantage of such a unit is its low-cost focus on an individual user, with no special room or setup needed. Within minutes of installation, a user can talk face to face to someone using another desktop system or someone with a very expensive room system. The disadvantage to this type of system is that it is not designed for conference rooms.

There are two common implementations of desktop video systems, PC based or specialized-appliance based.

PC-based Videoconferencing Terminal

A PC-based videoconferencing terminal is completely integrated into a PC. These systems are often based on a Windows platform with the necessary hardware and software components added. The primary advantage is that these systems can give the user immediate access to PC display capabilities in order to show PowerPoint slides or other software screens. These units typically support a method of application sharing as defined by the T.120 application-sharing standard. This allows an end user to collaborate with an endpoint by electronically sharing a document or whiteboard and sharing control of the document or whiteboard. The main disadvantage of a PC-based system is that it operates on PC operating systems, all of which can lock up or crash. Another disadvantage to a PC-based system is that not all multipoint conference units support T.120 application sharing, thereby limiting application sharing to point-to-point calls.

NetMeeting and CUseeMe are two examples of software-based H.323 clients. Software-based clients are often cheap to implement due to the low cost of simple USB cameras and cheap microphones. As well, the software is often free, which makes a software-based solution quite appealing to organizations with little or no funding for videoconferencing. The caveat is that software clients require very powerful systems to function well, and some clients don't work properly in conjunction with other H.323 systems. Software clients use the main system CPU to encode and decode video. This causes a great burden on the system, often causing choppy video or other problems.

Web clients offer the hope of simplicity to end users of videoconferencing. The idea of simply pointing the web browser to a site and almost instantly being able to connect to another user via the browser is a compelling and simple idea.

The clients are usually fairly easy to install. Assuming you have a working PC camera installed on your computer, the ActiveX controls or plug-ins detect the camera and configure things fairly automatically. However, these products are generally not standards based and they can only talk to other instances of the same software.

The best use of these products may be to create online groups. Online support groups, user groups, group chats, etc. can be supported by the products (although most currently seem to be point to point). Online technical support is another promising application, useful in helpdesks and other support functions. Adult uses are also one main use of these technologies at the current time. In the R&E environment, the only appropriate use of these technologies might be to support student–faculty meetings or student group interactions in distance learning environments.

Appliance-based Videoconferencing Terminal

An appliance-based videoconferencing terminal does not require a personal computer to function, being self-contained with proprietary hardware inside. Units contain their operating software in an internal flash-type memory bank, measuring boot times in seconds. Because these units are designed from the ground up to be videoconference units, the manufacturers do not need to make compromises to accommodate a Windows type of operating environment. User navigation menus are highly graphical with multiple levels of on-screen help available. The user interfaces are highly intuitive and are adapted to easily by the novice. A main advantage is ease of installation. With a PC-based system, the installer needs to possess the basic skills necessary for PC installation in a network environment. For an appliance-based unit, the user typically needs to hook it up to one or two video monitors, supply a network connection, power up the unit, and follow the on-screen instructions for software setup. A key drawback of these units is their lack of T.120 application-sharing support. Also, the lack of modularity means that when they fail, the entire unit needs to be sent in for service, rather than a small component of the system.

The primary advantage of desktop video products is they allow each person to have a personal video conferencing system in their office or home, allowing impromptu casual usage. Most of the desktop systems utilize your existing PC's or MAC's CPU and monitor, and rely on a network or phone connection to the other systems to offer visual conversations. Some of these systems also offer shared whiteboard and application sharing as well. Shared whiteboard gives each person the ability to simultaneously view and annotate a particular document, much the same way you would in a conference room having a fixed whiteboard. Application sharing allows multiple people to share documents and spreadsheets, interactively, using popular office applications. This eliminates the need to pass files back and forth for review and editing among workgroup members.

Supporting Video in the Converged Network

Businesses and end users developed converged networks for a variety of reasons. They were developed as a way to control costs by deploying one common network as opposed to separate, dedicated networks. Another reason was to facilitate new applications that took advantage of the multiple information types available to a given endpoint through a converged network. Integrating video into the converged network gives both of those advantages.

- Cost savings through the elimination of dedicated video networks
- New applications that combine voice, data, and video, like video contact centers, video messaging, business kiosks that provide data access with video interfaces, and many others.

Converged Network Components for Video

The ITU H.32x family of standards handles multimedia communications. This family includes H.320 (communication over ISDN [integrated services digital networks]) and H.324 (communication over traditional phone services).

H.323 is a communication standard produced by the ITU, initiated in late 1996, and aimed at the emerging area of multimedia communication over LANs (local area networks). It is an outgrowth of the traditional H.320 technology but optimized instead for the Internet. H.323 has since been revised to include voice-over IP and IP telephony, as well as gatekeeper-to-gatekeeper communications and other data communications that involve packet-based networks. These networks include IP-based networks like the Internet, Internet Packet Exchange (IPX) LANs, and WANs. H.323 is widely supported by many commercial vendors and used throughout the world in commercial and educational markets.

The H.323 standard specifies a great deal of information about the properties and components that interact within an H.323 environment. It specifies the pieces that combine to provide a complete communication service, as follows:

- terminals, either PC or stand-alone devices, are the endpoints of the communication lines
- gatekeepers, the brains of the network; provide services such as addressing/identification, authorization, and bandwidth management
- gateways, which serve as translators when connecting to a dissimilar network (such as an H.324, for example)
- MCUs (multipoint control units), which allow multipoint conferencing, or communication between more than two parties at once (much like a traditional conference call on a telephone)

In addition to component types, H.323 also describes protocol standards, permissible audio and video codecs, RAS (registration, admission, and status), call signaling, and control signaling. H.323 specifies a mandatory level of compliance and support for these specifications for all terminals on the network

The terminals and MCUs in a converged network are similar to what was discussed in previous sections. The major difference is that they use packet-based interfaces as opposed to traditional circuit-based interfaces. In fact, many vendors offer terminals and MCUs that support both H.320 and H.323 interfaces.

Figure 7.2 shows a typical IP-based video network.

Gatekeepers

An H.323 gatekeeper is assigned control of a particular set of videoconferencing resources such as terminals, gateways, and MCUs. In this role, the gatekeeper can provide or facilitate several services that enable H.323 conferencing to be more reliable and more secure. If a gatekeeper is present on the network, the H.323 standard requires that H.323-compliant terminals register themselves with the gatekeeper and allow the gatekeeper to identify them to others and control their activities within the zone. If a gatekeeper is not present, the standard allows for the terminal to control its own calls, placing them via IP address with no gatekeeper registration or intervention required. Once a terminal is reg-

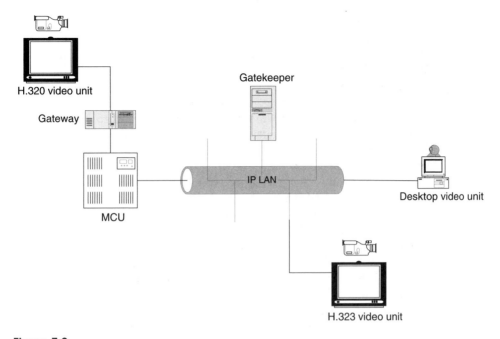

Figure 7.2

H.323 Video Network.

istered with a gatekeeper, the H.323 standard identifies some broadly defined key services that the gatekeeper could provide.

- Address translation—This function maps an alias or "video telephone number" of a user to the physical IP address of a terminal. This allows for people to call each other using user-friendly identification, such as a short numeric extension or an email address. Notably, a common schema for scaleable global addressing has not been defined.
- Admissions control—This function accepts or declines a call based on a variety of criteria, including available network bandwidth or specific user authorization level. Simple gatekeepers allow all calls through. (This level of call control is distinct from control at the terminal, where the end user can decide whether or not to answer any given call.)
- Bandwidth control and management—The gatekeeper can accept or deny calls based on the total available network bandwidth or based on a preset maximum number of simultaneous calls. This keeps videoconferencing calls from overloading the network. The gatekeeper may also handle requests from terminals for additional bandwidth during a call. In many ways, the bandwidth control and management functionality overlaps with the "bandwidth broker" and "policy broker" functionality under investigation as part of IP QoS (Quality of Service) development.

- Zone management—Each gatekeeper sets up a zone that may include terminals, gateways, and/or MCUs. The gatekeeper controls the identification of and communication between devices in its local zone, allows devices to join or leave the zone, and controls access to the local devices from H.323 devices outside the zone.
- Call control signaling—The gatekeeper can process call control signals for particular calls, or allow this information to bypass it and go directly to participating terminals. If the gatekeeper remains instrumental in call control, enhanced management and error handling are possible but with the trade-off of additional network and processing overhead.
- Call authorization—The gatekeeper can reject calls sent to terminals in its zone. The gatekeeper can also control what call types and resources are authorized for specific terminals. However, authentication is currently based on IP address and/or alias and not tied to any user-specific authentication.
- Call management and tracking—The gatekeeper can track current calls, log calls placed over time, and provide this call tracking information to other devices. Such information can be used for system administration and maintenance as well as for billing purposes.
- PBX functions—The gatekeeper can provide "PBX-like" services such as call identification, call forwarding, and call transfer. These features, in turn, can make possible applications such as a "video receptionist" and "video voicemail."

Gatekeepers today are available as full-featured stand-alone software applications or often the gatekeeper is co-resident with the MCU. In some newer, converged networks the video gatekeeper functionality is combined with the voice media controller.

Gateways

A gateway provides transcoding services such as address translation, network protocol translation, and audio/video coding translation between dissimilar conferencing technologies. This allows a user using one type of service such as H.323, for example, to connect to and communicate with another user that is using perhaps a radically different type of service. This potential ability not only provides a bridge between differing technology that might be present at any given time but also technology as it has changed over time, extending the useful life of technology that was built on previous standards.

One of the most common gateways for use with H.323 is designed to transcode between H.320 videoconferencing with its ISDN transport and H.323 as it travels over IP. Because many campuses have already invested heavily in H.320 and have mature H.320 applications, it is advisable to consider a model for H.323 deployment that includes such gateways as a complement to the IP-based service. This could also permit existing and future communications with areas that do not have high-performance IP networks available or where ISDN may be a more affordable option. A secondary use for the H.320-to-H.323 gateway could be to provide redundancy for a LAN-based MCU service. Should a network break occur, a conference could be routed alternately from one MCU, across a local LAN, through an H.320 gateway over ISDN, back through a second gateway, and onto the LAN local to the second MCU.

There are also common communication scenarios that call for the inclusion of traditional voice calls over the PSTN (public switched telephone network) in an H.323 communication. This may be as simple as needing to communicate with someone who has not yet implemented H.323 to extending H.323 services to users when they are mobile (e.g., using a cellular phone.) To enable this type of interconnection, an H.323/VoIP/voice gateway may be deployed. Other gateways—H.320-to-H.321 (ATM), H.323-to-H.321, and H.323-to-VRVS (Virtual Room Video Service, http://www.vrvs.org)—also exist and should be included where needed.

Because gateways function between protocols and not within a single protocol, some special configuration may be required. In particular, the RAS (registration, admission, and status) section of the H.323 specification, which permits dynamic conference ID registration, has no functional equivalent in the H.320 specification. The result is that if a gatekeeper is present, the conference ID must be pre-defined for multipoint calls. Point-to-point calls not using a gatekeeper do not require special treatment.

A second configuration issue to be careful of is that IVR (interactive voice response) systems often use the asterisk ("*") to signal request for operator. In such an environment, predefined groups intended for use with gateways shouldn't include asterisks. Unfortunately, this requirement conflicts with the trend among H.323-only users to utilize the asterisk as a delimiter.

Some CPU-intensive audio transcoding can cause significantly delayed audio, resulting in an objectionable lack of audio/video synchronization. H.323 systems use G.723 and G.711 while H.320 systems use G.728 and G.711. G.711, the protocol in common, provides toll-quality audio but uses 64 Kbps. Disabling transcoding minimizes the audio delay due to the transcoding but would leave only 64 Kbps available for video in a 128 Kbps single-circuit ISDN call. Enabling G.728–G.711 transcoding would reduce the audio bandwidth requirement to 16 Kbps and free an additional 40 Kbps for video. In a 384 Kbps triple circuit bonded ISDN call, minimizing the audio delay might be deemed worth the minimal video degradation. Whether or not to permit audio transcoding should be decided on a call-by-call basis.

Supporting gateways can be operationally complex. Some service providers have recommended that users implement dual-technology codecs. For example, many group conferencing systems support H.323 and H.320. By recommending these systems, a service provider can centrally support the more desirable H.323 IP-based technology while allowing the user to manage (and pay for) their own dedicated ISDN BRI lines for H.320 compatibility. This has the benefit of ensuring that the user has access to legacy technology if needed, but at a cost that encourages migration to more modern protocols.

In most videoconferencing over IP, including H.323, the endpoints compress the audio and video data to be exchanged so that transmission of that data over the network will be faster and less likely to degrade, and will consume less network bandwidth. The trade-off for this added network efficiency is a need to compress the media stream as it leaves the sending endpoint and then decompress the media (audio/video) stream for interpretation once it arrives at the receiving endpoint. This compression and decompression require processing power and additional time. The less processing power there is, the more delay is introduced into the communication. H.323 videoconferencing is often considered a desktop technology but PCs in general are not powerful enough yet to do high-quality full-screen, full-motion video compression and decompression, especially in addition to all the other applications that a PC might be running at the same time a videoconference is desired. This has forced leading H.323 vendors to integrate hardware-based codecs ("compressor/decompressor") into their videoconferencing products. These codecs are specifically designed to offload the compression and decompression task from the PC, allowing the endpoint overall to achieve good performance. In the past and often still today, the codec is included as an additional PCI bus card in the PC. Some PC products can support a high frame rate (15–30 frames per second) and extended range of call quality/bandwidth settings (128 K–1.5 Mb) so that videoconference quality seen at one's desktop can equal that of larger and more expensive room-based conferencing systems.

More recently, with the advent of USB (Universal Serial Bus) technology, the trend is more toward external devices that plug into a USB port. This allows for the simplicity of "plug and play" installation for seemingly simple videoconferencing devices that are also capable of providing relatively high call quality. The extra processing power required is included in the USB camera/device, with the USB port providing the "bandwidth" necessary for the compressed video to pass from the camera/device to the PC.

Another addition to the "hardware line-up" of videoconferencing endpoints is standalone non-PC-based appliances. These appliances are specialized hardware devices (system/camera combinations) that provide high-quality medium- to large-scale videoconferencing capability. They do not run other programs such as a PC-based endpoint might and they are larger and more expensive, yet often as simple to use, as USB devices on a desktop PC.

Overall, though hardware-based endpoints cost more than their software-only counterparts (which range from inexpensive to free and are discussed in the following section), the extra cost is often justified in order to achieve videoconferencing call quality that is acceptable beyond just novelty or casual use.

Video and QoS

Videoconferencing was originally deployed over networks that could provide some guarantees about the level of service that would be delivered to the application. The ISDN and/or dedicated T1 circuits of the H.320 standards-based world provided predictable delays over dedicated paths. This allowed videoconferencing vendors to create products to work within these parameters. However, dedicated circuits are also expensive circuits.

H.323 standards-based videoconferencing was engineered for videoconferencing that takes place on a data network without any quality-of-service standard, such as the Internet. Such networks were not originally intended for delivery of sensitive near real-time applications. The data network is used for multiple purposes: e-mail, web browsing, and other activities are inter-mixed with H.323 videoconferencing.

The audio/video information within a videoconference is segmented into chunks by the application, encoded and compressed, put into a series of data packets, and sent over the network to the remote end at basically constant intervals. The data packets may arrive at their destination at slightly varying times, if at all, and possibly out of order. To keep the "real time" impression of an interactive videoconference, the packets must arrive on time and in time to be re-ordered for delivery through the videoconferencing terminal.

Video Transmission Requirements

There are five fundamental network problems for videoconferencing over packet-based networks. They are *bandwidth, packet loss, latency, jitter, and policies.*

Bandwidth is the fundamental requirement that there be enough space in a network path for all of your packets to get through unimpeded. A typical ISDN videoconference uses around 128–384 kb/s. IP-based H.323 video systems can use the same bandwidth, although in general they tend to go higher because the network is less expensive, so bandwidth of around 384–768 kb/s is very common. Streaming video applications and higher-quality videoconferences can go to 1.5–2.0 Mb/s, and if you want to go to broadcast quality the sky is the limit—6 Mb/s for NTSC/PAL transmission, 20 Mb/s for prerecorded HDTV, and higher for "live" content.

This bandwidth need is symmetric—each end will transmit and receive this amount of traffic. If you are in a multi-point videoconference, then you need to keep in mind that the MCU/bridge is seeing all of the streams at the same time, even if it is not forwarding them on. So if you have an 8-site videoconference running at 384 kb/s, every site sends and receives 384 kb/s to the MCU, and the MCU receives and forwards 8*384 kb/s or 3 Mb/s.

Packet loss is when packets fail to arrive correctly. This can be due to insufficient bandwidth along the path (when congestion occurs, routers will drop packets), or perhaps errors in transmission. Errors occur most commonly on wireless links such as microwave, satellite, or local wireless Ethernet. They can also occur on copper and even fiber links. Packet loss results in effects such as "tiling" within the video window, missing pieces or blank areas within the video window, and/or disruptions in audio.

Video communication is even more sensitive to packet loss than voice communication. Industry guidelines for packet loss typically show the following:

- A 1 percent packet loss may produce blocky video and/or audio loss
- A 2 percent packet loss may make video unusable, although audio may sound somewhat acceptable
- Packet loss above 2 percent is unacceptable for H.323,

Latency is the time delay between an event occurring and the remote end seeing it. Latency is introduced both by the encoding/decoding process (and hence depends on the equipment used), and also by the time it takes packets to traverse the network. There is little you can usually do to change the network latency on any large scale, beyond getting directly involved with a carrier or a research network. The speed of light is a limiting factor, especially on satellite networks or international cable links.

Excessive latency increases the chances of people "talking over one another" because they don't realize that the person at the other end has started speaking too. Another problem is that the latency for the audio and video may be different, and hence lip movements don't appear synchronized with the audio. This is a function of both the terminal and the network, and can vary dramatically— some products try to compensate for it.

Latency guidelines for good video quality are very similar to those for voice. These guidelines typically are

- Under 150–200 ms delay can give very good video quality.
- Delays exceeding 200 ms may still be quite acceptable depending on customer expectations, codec type, etc.

- The H.323 protocol defines a maximum end-to-end delay of 400 ms. Delays beyond this level can cause network instability.
  (These numbers are for delay between endpoints, meaning LAN/WAN measurements not including IP phones.)

Jitter is the random variation in latency, due to things like other processes running on the terminal (for example on your desktop PC), other traffic temporarily blocking the path through the routers along the way, or even the network path changing during a videoconference. In extreme cases this results in packets arriving out of order from their transmitted order. Jitter results in uneven and unpredictable quality within a videoconference. Terminals will try to compensate for this by buffering the traffic up to some finite time, before playing it out. This can increase the latency even further.

Policies are introduced by things such as firewalls and network address translation (NAT) devices that are generally used to try to hide or protect network elements from the wider Internet. H.323 uses dynamically allocated ports, usually, and is thus not very firewall-friendly. Unfortunately there are very few technical solutions to these, and they require you to discuss policy issues with your, or their, network managers.

Network Address Translation (NAT) and Firewalls

Videoconferencing is often recognized as a difficult service to negotiate with Network Address Translation (NAT) and Firewalls. The H.323 standard defines a service that utilizes bi-directional communication for call setup with connections characterized by dynamic port utilization and a high data rate. There are both UDP and TCP parallel connections made on the dynamically negotiated ports. The situation is further complicated by the transmittal of this port information in the payload section of the IP data stream that makes up this negotiation (RFC 2663). The difficulty in the deployment of videoconferencing applications is due to the complexity of the H.323 protocol and its dependence on some of the utilized components (i.e., gatekeepers, MCUs, etc.) and the variety of vendor implementations.

NAT and Firewall technologies were deployed to solve important security issues by limiting access to an internal network's ports by filtering inbound Internet traffic. In addition NAT provides IP address space to the internal network by brokering a single port to multiple internal ports. Because of the widespread use of NAT and Firewalls, and given the characteristics of videoconferencing systems, these particular network components and their configurations have become an obstacle to deploying H.323 application systems. NAT is actually a form of firewall but is so widely deployed in home DSL/Cable routers and 802.11 Wireless Access Points that it deserves individual consideration.

All TCP/IP applications depend on network routers to control the direction of IP packets to ensure that each packet reaches its intended destination. Routers that function in their purest form do not pose a problem to H.323 traffic in particular. What frequently causes problems with H.323 traffic is a mechanism to conserve IP addressing space called Network Address Translation, or NAT. NAT accomplishes this by transparently sharing a single IP address with multiple hosts. By translating the IP address used on a private network to an IP address suitable for the public Internet, the NAT-enabled router can map private IP addresses and private ports to external IP addresses and external ports and therefore support multiple private IP addresses. The important aspect to remember of the relationship between the H.323 application and the network router is that the application is unaware that traffic is running through a NAT facility. Another problem exists because NATs typically set up their port mappings by examining the applications packet header. While this works well for most applications, the H.323 standard requires IP address and port information to be stored in the data portion of the IP packet.

Most NAT implementations will allow for the configuration of a DMZ host or, in other words, to pass all inbound traffic to a particular internal host. This is not always without some packet modification that would preclude H.323 communication, and certainly introduces significant security vulnerabilities into an internal system. A more acceptable alternative is an application layer gateway or proxy that is able to interpret the call setup traffic and create the required port configuration. Most clients support proxies but do require address configuration.

As stated earlier, NAT is a general type of firewall technology. The other main types are packet filtering, circuit gateway, and application proxy. Any discussion of firewalls and their various implementations and deployment strategies is essentially eternal, so this material should be considered superficial to the extreme. Consideration must also be given that firewall configurations commonly require asynchronous profiles for internal and external clients and frequently videoconferencing will transverse multiple firewalls.

Packet-filtering firewalls block or allow connections based entirely on addressing information in the IP header. As we discussed earlier, H.323 embeds routing information in the actual data or payload of the packet that responds to a signaling request, so packet-filtering provides no method of associating the UDP request with the routing information. Consequently, the only way for a packet-filtering router to support H.323 is to open up all UDP and TCP ports above 1024 in each direction. Obviously, both source and destination firewalls would have to be configured similarly, therefore significantly reducing the protection the firewall was implemented to provide.

A circuit-gateway firewall allows a UDP request to initiate the opening of dynamic ports for a limited time to allow for streams associated with the application. Most firewalls utilizing this method have some understanding of common application protocols such as Telnet and FTP. Their behavior is predictable because of static port requirements and the result is an expected state. The circuit-gateway firewall can provide adequate support for videoconferencing if it can disassemble the payload and respond by opening the requested ports dynamically. The disassembling of the packets and associating the contents with the UDP signaling request is complex, requires more than standard application savvy, and can introduce latency.

An application-proxy firewall is evident to the application because it implements a partial H.323 stack. The proxy performs address translation but can manage the internal and external ports by actually participating in the signaling process and thereby opening the requested data ports. The proxy firewall behaves transparently to both internal and external clients after the connection is made, but making that connection is not without cost. Because the clients are aware of the proxy, they must be configured to both place and receive calls through the proxy IP address. The clients must be configured with this address (like any proxy) and the H.323 identity fields must be completed to enable access.

Of course one solution to supporting videoconferencing across the firewall would be to merely open the required ports to all network traffic. Many vendors have now implemented firewall technologies into router products that acknowledge the limitations of older implementations and properly support H.323 and SIP network traffic. These solutions can detect the videoconferencing signaling requests and take appropriate action to allow the traffic to transverse the router or firewall.

Port Forwarding

One piece of information contained in the packet header is something called the port number. Source and destination port numbers, combined with source and destination addresses, uniquely identify a connection. Port numbers often refer to applications or services.

This connection identity method must, obviously, change somewhat where NATs are used. Much as how the gateway or router remaps unregistered, private IP addresses to registered, public IP addresses, it also remaps the port numbers. This is done using a port-mapping table to remember how the ports were renumbered for outgoing packets. The router can also reverse this process so that returning packets reach the correct incoming destination.

Modern firewalls and routers can be configured to pass contacts through to certain sites and certain ports. For example, if we want to provide a service to the outside world, say an anonymous ftp service, we can instruct our firewall to allow packets through on port 20/21 if they contain our server's destination address. In the case of videoconferencing through a private network, we simply tell the router which port numbers to forward (basically, which application's packets to allow through).

Protocols for Supporting QoS for Video

The QoS tools available for video are the same as those for voice and are discussed in detail in Chapters 6 and 13. Some of them are 802.1 p/Q, Differential Services Code Point (DSCP), Resource Reservation Protocol (RSVP), Priority Queuing (PQ), and Port Numbers.

Resource Reservation Protocol (RSVP) is being investigated much more for video support even than for voice support. Because of video's increased bandwidth support needs, reactive mechanisms like WFQ or CBQ may prove insufficient to support high-bandwidth, real-time services. In the real world, converged voice, data, and video networks will most likely need to support all of these protocols, including both prioritized queuing and RSVP.

Video and the Wireless Network

Video on the Wireless LAN

Wireless local area networks are now becoming more and more common, and so the question arises as to whether videoconferencing can be done over such networks. Video and audio are more demanding of network quality than are simpler things such as email and web browsing. Dropped packets and network traffic competition cause video and audio to fail long before email and web browsing. The net result is that the wireless distance that can be achieved for video is significantly less than that for other applications.

Another factor that works against wireless video is that wireless networks are completely shared by all users, as if dumb hubs were being used on a wired network. Most wired networks today use switches, which isolate the traffic from one station to another. Thus if there are many users on wireless network, then video will degrade faster than other applications.

In Chapter 8 we will discuss wireless networks in more detail. Video over wireless networks is supported very much like voice over wireless with many of the same capabilities and issues.

The predominant protocol used to support Wireless Local Area Networks (WLAN) is the 802.11(x) series of protocols. 802.11b provides WLAN support with up to 11 Mbps of bandwidth on a 2.4 Ghz radio frequency. 802.11b is often known as WiFi (wireless fidelity). 802.11a provides up to 54 Mbps on a 5 Ghz frequency. 802.11g is a new, proposed standard for providing the bandwidth of 802.11a on the 802.11b frequency.

Because of the bandwidth needed to provide video services on the converged network, it is believed that video applications will be one of the major driving forces behind the deployment of the higher bandwidth wireless standards like 802.11a and 802.11g.

In Chapter 8 we will discuss in much more detail some of the issues encountered in deploying real-time services such as voice and video over the WLAN. In summary, the main issues are

- The 802.11(x) protocols do not currently have any mechanism for supporting Quality of Service. Thus there is no way to prioritize real-time services. This is being addressed by the proposed 802.11e standard.
- Because WLAN are broadcast media, there can be serious security issues associated with WLAN deployment.

Again in Chapter 8 we will discuss methods to overcome these issues that apply equally well to voice as well as video.

As standards like 802.11a, 802.11g and 802.11e come into production environments, it is likely that there will be a large increase in the deployment of video applications over WLANs similar to the video applications being deployed on wired LANs.

Video on Public Wireless Services

One of the major sources of excitement in the video networking industry is the advent of video services on public, wireless services. These are being deployed via satellites and, more significantly, over public wireless carriers.

Satellite

Videoconferencing has been done via earth satellites for many years, with significant success. But it suffers from the same effects as wireless video, and a few more. The latency is inherently greater due to the speed of light delay to and from the satellite. This is at least 1/2 second. However, most users quickly adjust to that effect and then ignore it. Other issues are that the satellite data rates may be different for uplink than for downlink, making symmetrical videoconferencing more difficult. The satellite may also operate in some form of store-and-forward transmission mode, making the jitter larger.

Because of these issues, satellite-based video applications tend to be in specialized niches like broadcast news and specialized international videoconference applications.

Public Carrier Wireless

Public wireless carriers are currently planning at least three different types of mobile video services: mobile video messaging, mobile video distribution services, and mobile video telephony/conferencing.

- Mobile video messaging is a person-to-person or person-to-machine communication service that consists of sending video content together with other media on a non-real-time basis from mobile to mobile, mobile to PC, or PC to mobile. The 3rd Generation Partnership Program (3GPP) has defined a non-real-time Multimedia Messaging Service (MMS) to allow mobile users to send and receive messages including a combination of one or more media elements.
- Mobile video distribution services allow mobile users to either stream or download video content to their mobile devices. The main difference between download and streaming is that the latter allows transferred data to be processed as a steady and continuous stream by a streaming multimedia application and displayed before the entire file has been transmitted.
- Video telephony is a person-to-person communication service that uses visual as well as voice data. It enables a real-time two-way stream of video and audio signals between two mobile devices or a mobile device and a fixed videophone.

It is important to understand the characteristics of each one of these mobile video services, as they impose different technical requirements both on the terminals and mobile networks. Although some of these services will be technically feasible after the introduction of GPRS, the more demanding mobile video services (such as mobile video telephony) will only be a reality in the long term.

Mobile equipment vendors have expanded their SMS (short message service) and WAP (wireless application platform) solutions to enable the delivery of rich content to mobile devices. The vendors' product roadmaps comprise enhancements to the vendors' SMS systems, to combine the conventional short messages with simple pixel pictures, and MMS, which will enable mobile video messaging applications. Some European operators have announced MMS trials with equipment providers.

Internet video content and streaming applications have experienced significant growth over the last few years. This increase in streaming usage is closely linked to huge media events such as the Olympics, World Cup, and national elections.

It is important to recognize, on the other hand, that the first video-enabled mobile terminals are already available and numerous media companies and vendors developing products and services are now focusing on the delivery of rich media to wireless terminals.

Despite some early product examples, handsets are still one of the biggest limitations for the development of the mobile video market, and will require a significant transformation from today's design and functionality. Some of the new features that will have to be integrated into future mobile devices in order to view video content include high-resolution color flat panel displays, client video players (either software or hardware) to be able to playback video content, additional memory for buffering streamed video and storing downloaded small video clips or multimedia messages. In addition to them, mobile devices capable of generating video content require integrated CMOS video cameras and video codecs.

Software video decoders represent a cheap and viable option for applications where the mobile device is used for playback only (i.e., mobile video messaging and mobile video distribution services). General-purpose Digital Signal Processors (DSPs) with enough computational power for decoding video signals are presently available on the market that integrate software decoders. Several companies currently offer free downloadable software video players for PDAs and other handheld devices. Some of them have also developed proprietary compression technology and thin client software applications optimized for devices with processing and memory constraints.

Hardware-based codecs will be preferred in mobile phones as they consume less power and are faster than software routines. Some hardware codecs are presently available in the market, but current products need to evolve in order to reduce cost and power requirements. However, significant increase in computational capacity is needed for software-based encoders. Therefore, they will most likely be used to record and play video (and audio) on PCs, using the computer's CPU for processing.

Most of the handset manufacturers will most likely launch video-enabled mobile phones as part of their initial 3 G product ranges, which will likely be available in mass quantities in 2003–2004. In general, these new terminals will include color LCDs, video download and video streaming capabilities (i.e., streaming video players, multimedia cards for external storage), and CMOS video or still image cameras. Before that, GPRS MMS phones for GPRS-supporting services based on photographic images only (i.e., no video capability) will start to appear in 2003.

3GPP standard compliant (and MPEG-4 based) video telephones will constitute a next generation of video-enabled phones.

Even though the range of mobile terminals will become much wider than it is today, not all future mobile terminals will be video enabled. The terminal replacement rate, as well as the penetration of data-related devices, is a key factor for understanding the mobile video services uptake. A slower acceptance of video-enabled devices would have a significant impact on the success and uptake of all mobile video services. Upcoming 2.5 G and 3 G mobile networks will enable mobile applications demanding higher data rates and more stringent quality of service requirements.

With the widespread introduction of GPRS technology, mobile video messaging services and downloading small video files into the mobile device will be possible due to the increased data rates provided and the non-real-time characteristics of these services.

Streaming video distribution services are best supported with 3 G. However, some VOD niche applications could also be designed for GPRS networks offering small low-resolution video files that require low data throughput. Streaming full-length movies to a phone or handheld will be, at best, a niche application for a long time.

Some mobile network technologies are not appropriate for real-time video telephony/conferencing. For instance, GPRS latency (approx >300 ms) and packet loss (>1%) characteristics are inadequate to make it a suitable technology for video telephony. 3 G is the first technology that can be widely used for video telephony services, as it offers circuit switched data rates of 64 Kbps both to and from the terminal.

One of the big stumbling bocks ahead of all video applications, especially in the public carrier space, is *are users ready to see and be seen?* It is envisioned that the introduction of video will create a change in users' communicational behavior. Visual applications will not necessarily replace other existing ways of communications. The mobile user will communicate using the most convenient communication vehicle at each time: the one that best suits his or her current situation and needs is available/feasible and is charged at a price that the user is ready to pay. From the users' perspective, price sensitivity will be one of the key issues for all these new mobile video services. Nevertheless, it is clear that mobile users will react differently to each of the mobile video services.

As we have discussed throughout this chapter, the next logical addition to the converged network is video communications. Video can take several different forms from simple video conferencing to high-quality streamed video applications. Because video traffic needs share many needs in common with voice traffic, many of those needs can be supported by the same techniques as those used for video. Video, however has some unique needs of its own, specifically bandwidth and the need to connect to multiple endpoints simultaneously.

Video can be supported on multiple types of converged infrastructure including both wired and wireless services. In the future, video-based applications will see considerable growth as these technologies mature.

CHAPTER 8

CONVERGENCE ON WIRELESS NETWORKS

One of the fastest growing areas in IP networks today is the proliferation of Wireless Local Area Networks (WLAN). As more and more workers need access to the IP data network from multiple locations and in highly mobile environments, removing the physical restraints on the IP network becomes very attractive.

In parallel with the increased need for mobile IP networks, many of these same users are beginning to deploy converged voice and data applications. Where there is a strong case for integrating voice and data on the same network and a need for mobility, voice over a WLAN becomes a compelling solution.

In this chapter we will begin by discussing the fundamentals of the wireless IP infrastructure. In the second section we will discuss supporting voice over the WLAN including QoS and security considerations. At the end of this chapter we will briefly discuss the newly emerging public wireless voice/data services.

Wireless IP Networks

One of the most confusing aspects of wireless LAN is the terminology used in discussing it. Terms like WLAN, WiFi, 802.11, wireless Ethernet, and so on are used almost interchangeably in some contexts. Wi-Fi is short for *wireless fidelity* and is another name for IEEE 802.11b. It is a trade term promulgated by the Wireless Ethernet Compatibility Alliance (WECA). "Wi-Fi" is used in place of 802.11b in the same way that "Ethernet" is used in place of IEEE 802.3. Products certified as Wi-Fi by WECA are interoperable with each other even if they are from different manufacturers. A user with a Wi-Fi product can use any brand of access point with any other brand of client hardware that is built to the Wi-Fi standard.

Wireless LAN Components

All Wireless LANs consist of a few common, basic components. The most basic component is an endpoint access device, typically a wireless network interface card (NIC) in a PC or the equivalent. These devices may be removable like a PCMCIA-based NIC or a USB-based PC interface, or may be built into the device hardware. The interface includes the transceivers used to send wireless packets and is controlled by software that supports the wireless LAN protocols.

In the most basic form of wireless LANs, all that is needed are two or more devices with WLAN interfaces which can send information between them wirelessly. This is sometimes referred to as Ad-Hoc wireless networks. (See Figure 8.1).

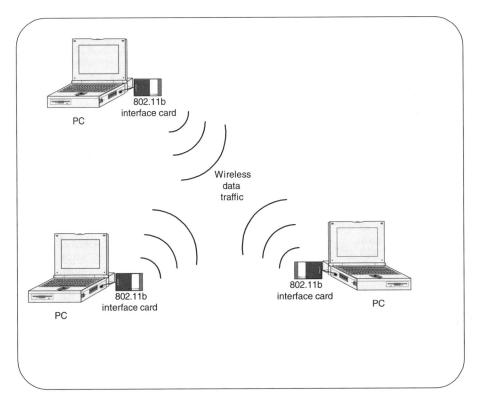

Figure 8.1

Ad-Hoc 802.11b network.

Larger and more complex networks need some form of central connection. This central connection is known as an *Access Point* (AP). The access point provides for a common point where WLAN endpoints can send and receive their data. In effect, the AP functions like a wireless Ethernet hub. Multiple APs can be connected together over a wired or wireless network to build a set of coverage areas in which a wireless endpoint can "roam" from one access point to another.

These coverage areas can also be set up to overlap, thus allowing a user to have continuous coverage over a large area (see Figure 8.2). This type of deployment is sometimes referred to as Infrastructure Mode.

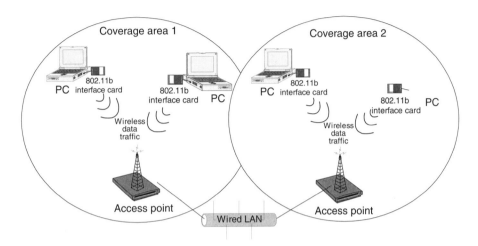

Figure 8.2

Infrastructure Mode 802.11b network.

There are also other, more specialized wireless LAN components. *Access Servers* are specialized forms of Access Points that provide for user registration, authentication, and billing services. *Point-to-Point Wireless links* are used for interconnecting multiple sites over a fixed wireless link. These links are often used to connect WLANs in multiple locations together or to provide links between LANs where no wired solution is available. Many specialized antennas are also available for extending the range of the WLAN or for use in specialized applications.

Wireless LAN Protocols

There are several protocols used in WLANs. By far the most common ones used are based on the IEEE 802.11 standard. There are several different versions of the 802.11 protocol: 802.1b, 802.1a, and a recent addition 802.1g.

802.11b

802.11b is an enhancement of the initial 802.11 Digital Spread Spectrum Service (DSSS) physical-layer specification includes 5.5 Mbps and 11 Mbps data rates in addition to the 1 Mbps and 2 Mbps data rates of the initial standard. 802.11 finalized this standard (IEEE Std. 802.11b-1999) in late 1999. To provide the higher data rates, 802.11b uses CCK (Complementary Code Keying), a modulation technique that makes efficient use of the radio spectrum.

Most wireless LAN installations today comply with 802.11b, which is also the basis for Wi-Fi certification from the Wireless Ethernet Compatibility Alliance (WECA). These products have been available for several years.

The most currently available products on the market are based on 802.11b. Because of this, most public and private wireless LAN services are 802.11b.

802.11a

802.11a is a Physical Layer standard (IEEE Std. 802.11a-1999) that specifies operating in the 5 GHz UNII band using orthogonal frequency division multiplexing (OFDM). 802.11a supports data rates ranging from 6 to 54 Mbps. 802.11a-based products became available in late 2001.

Because of operation in the 5 GHz bands, 802.11a offers much less potential for radio frequency (RF) interference than other versions (e.g., 802.11b and 802.11g) that utilize 2.4 GHz frequencies. With high data rates and relatively little interference, 802.11a does a great job of supporting multimedia applications and densely populated user environments. This makes 802.11a an excellent long-term solution for satisfying current and future requirements.

802.11g

802.11g is a higher speed extension (up to 54 Mbps) to the 802.11b physical layer specification, while operating in the 2.4 GHz band. 802.11g will implement all mandatory elements of the IEEE 802.11b standard. For example, an 802.11b user will be able to associate with an 802.11b access point and operate at data rates up to 11 Mbps. In early 2002, 802.11g decided to use OFDM instead of DSSS as the basis for providing the higher data rate extensions.

An issue is that the presence of an 802.11b user on an 802.11g network will require the use of RTS / CTS (request-to-send / clear-to-send), which generates substantial overhead and lowers throughput significantly for all 802.11b and 802.11g users. RTS / CTS ensures that the sending station first transmits a RTS frame and receives a CTS frame from the access point before sending data. A mixture of 802.11b and 802.11g requires RTS / CTS to avoid collisions because 802.11b stations can't hear 802.11g stations using OFDM.

The 802.11g standard is still under development, with a final standard likely available by the end of 2003. With pre-standard chipsets just becoming available now, product vendors will probably release 802.11g radio cards and access points in early 2003.

802.11a vs. 802.11g

As the need for higher data rates in wireless networks increases, there is currently a debate in the industry on which standard for 54 Mbps wireless will dominate.

The 802.11a standard and FCC spectrum regulatory status is firmly in place. Chipsets have been available for nearly a year, and several product vendors are now shipping 802.11a access points and radio NICs. This places 802.11a ahead of 802.11g in the market by about six months. A big difference with 802.11a is that it operates in the 5 GHz frequency band with twelve separate non-overlapping channels. As a result, you can have up to twelve access points set to different channels in the same area without them interfering with each other. This makes access point channel assignment much easier and significantly increases the throughput that the wireless LAN can deliver within a given area. In addition, RF interference is much less likely because of the less-crowded 5 GHz band.

Similar to 802.11g, 802.11a delivers up to 54 Mbps, with extensions to even higher possible data rates by combining channels. Due to higher frequency, however, the range (around 80 feet) is somewhat less than in lower-frequency systems (i.e., 802.11b and 802.11g). This increases the cost of the overall system because it requires a greater number of access points, but the shorter range enables a much greater capacity in smaller areas via a higher degree of channel reuse.

A huge problem with 802.11a is that it's not directly compatible with 802.11b or 802.11g networks. In other words, a user equipped with an 802.11b or 802.11g radio card will not be able to interface directly to an 802.11a access point. In applications where there is little or no control over what radio NICs are used, there can be interoperability issues. The cure for this will come eventually, however, as multimode NICs become the norm.

Both 802.11a and 802.11g operate at 54 Mbps using OFDM. 802.11a provides greater total capacity and is less likely to encounter RF interference. It's relatively easy and cost effective, however, to migrate from an installed 802.11b network to 802.11g and maintain a high degree of interoperability. The multimode 802.11 NICs will eventually help eliminate the interoperability issues, though at about the time you will be able to purchase fully-compliant 802.11g access points.

The following are guidelines to help decide which of these two protocols may be more suitable.

Choose 802.11a if you're implementing a wireless LAN from scratch and need to install it today. You can purchase 802.11a products now, but 802.11g products won't be available for another six months or so.

- Consider waiting for 802.11g products if you have a relatively large installed base of 802.11b. You'll be able to upgrade to 802.11g rather easily.
- If you don't have high-performance needs today but might in the future, consider deploying dual-slot access points. That will enable you to install 802.11b access points now, and you can add an 802.11a NIC later. Of course your users will need to have 802.11a/b NICs to ensure interoperability.
- Implement 802.11a now within existing 802.11b networks if you have specific areas needing high performance (e.g., conference and computer rooms). As with the preceding point, however, end users will need 802.11a/b NICs to ensure interoperability.

Keep in mind that you should base the decision for implementing either 802.11a or 802.11g on existing or future needs for higher performance. Many applications can get by with what 802.11b offers, so be sure that you need the extra bandwidth. If a year or two from now you need higher performance, then you can easily upgrade to 802.11g or plug-in 802.11a where needed.

Supporting Voice on Wireless IP

The largest and most effective opportunities for WLAN-based telephony are in the corporate enterprise. The corporate world has been the first area to embrace Wireless LANs. These businesses can control coverage area, bandwidth utilization, and Quality of Service (QoS) implementations. Their users also have access to the corporate telephone system, giving wireless users the same features and accessibility as their wired peers. It can be cost effective and efficient to leverage the investment in wireless LAN by adding wireless telephones.

Most wireless users are those who are not usually tied to a single desk. Examples of this type of user are nurses, teachers, store clerks, IT managers, and executives. These types of users can generally be broken down into two categories: decision makers who need to always be available for quick responses, and those who are carrying out tasks that require them to be mobile throughout the workplace.

Vertical markets such as education, healthcare, retail, manufacturing, and warehousing were early adopters of wireless LAN technology. Employees in these industries are more mobile than the average office worker and have specific application needs that lend themselves well to handheld devices. Adding WLAN-based telephony leverages existing investments in wireless LANs while increasing productivity and responsiveness for mobile employees in the workplace.

Another area of significant interest and opportunity is the wireless public access market. In this environment, wireless users have connectivity in so-called "hot spots"—hotels, airports, coffee shops, convention centers, train stations, and the like. The target data user is the "road warrior" who travels frequently outside the office. The user's data applications are often served by public Internet connectivity, and his or her voice communication needs are served by public cellular access. In contrast, the target WLAN-based voice user for access networks is more likely the mobile employee at the site. The employee at the hotel, for example, can realize the benefits of mobile accessibility through the hotel's telephone system while sharing the same WLAN network, and without any airtime or usage charges.

The SOHO and residential markets are often touted as the new frontier and land of opportunity for WLAN networks. Wireless data applications from any room in the house offer appealing new applications. For the near term, however, ubiquitous telephone services from the local telephone carrier and inexpensive cordless phones will continue to dominate the SOHO and residential markets. When the cost of WLAN telephones can match what is already available at the local superstore, and low-cost residential IP telephone service is readily available, WLAN telephony will emerge strong in the home.

The devices used for WLAN telephony vary, depending on user needs. Some users require application-specific clients that focus primarily on voice, with little or no data capabilities. Voice-only users need simple-to-use mobile telephones with high-quality voice and advanced telephony features. Voice-only devices may be designed for a specific vertical market focus, such as high durability or hands-free operation.

Other users require combined voice/data access devices such as a wirelessly connected PC with VoIP applications that can use the combined access to voice and data services.

Multi-use devices such as PDAs that offer WLAN voice applications are another viable option for the mobile user that needs access to both voice and data applications from a small, portable device. Wireless PDAs with voice functionality usually offer lower-quality audio and less usability in exchange for a single device supporting many applications.

Voice as a wireless IP application presents unique challenges for WLAN networks. Primary among these is acceptable audio quality resulting from minimized network delay in a mixed voice and data environment. Ethernet, wired or wireless, was not designed for real-time streaming media or guaranteed packet delivery. Congestion on the wireless network, without traffic differentiation, can quickly render voice unusable. QoS measures must be taken to ensure voice packet delays stay at acceptable levels. Task Group E of the 802.11 committee is working on a standard for QoS for voice and multimedia applications. The promise of the 802.11e standard is that real-time applications such as voice and streaming video will be guaranteed packet delivery within acceptable limits. In the meantime, proprietary QoS methods are being implemented by several manufacturers to support wireless voice applications.

A second unique consideration for voice is mobility. Telephone users are simply more mobile than data users. It is easier to walk and talk than to walk and type. Wireless telephone users are likely to roam between access points more frequently, requiring seamless, low-latency handoff between access points. Supporting WLAN telephony may require adding wireless coverage into areas that do not require coverage for data applications, such as stairs, hallways, and outdoor areas.

Third is the need for security on WLAN networks. To ensure network privacy, all WLAN devices need to take additional measures to avoid intrusion. The unique challenge for voice applications is to provide added security without compromising voice quality due to delays or interruptions while initiating a call or roaming between access points.

QoS in Wireless IP

As currently implemented, the 802.11 suite of protocols does not support any method for differentiating types of traffic, and these protocols provide for no means of prioritizing one packet over another. This means that there is currently no method for providing QoS on a WLAN. This leaves any voice services over a WLAN vulnerable to long delays in congested networks.

Right now, there are two proposed ways to overcome the QoS issue on a wireless LAN. The methods are setting up separate voice and data WLANs or a new, proposed QoS protocol for WLANs called 802.11e.

A few vendors have developed proprietary QoS protocols for voice on the WLAN, most notably Spectralink's SVP protocol. Spectralink SVP provides a proprietary method of encapsulating the voice packets and software for the access points to provide the prioritization of the voice packets.

Separate Voice and Data WLANs

Because no current non-proprietary method exists for prioritization on the wireless LAN, many users are implementing a relatively simple, interim approach to providing good voice quality on the wireless LAN. This approach takes advantage of the recent trend in cost reductions for 802.11-based hardware.

What a lot of users are setting up is essentially two parallel wireless LANs, one for the regular data services and one for voice over a wireless LAN (see

Figure 8.3). Currently there are even access points on the market that support two radios and can support just this type of configuration. The two radios are set up as separate VLANs on the wired network.

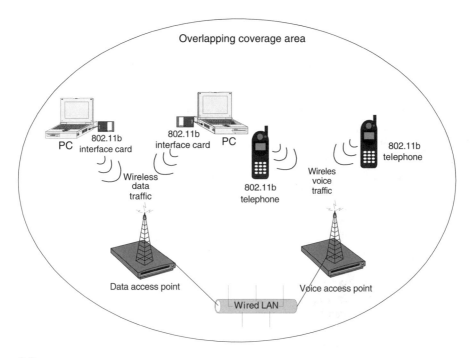

Figure 8.3

Separate voice and data WLANs.

There are two obvious drawbacks to this solution. The first is twice as much equipment as needed at twice the expense. The second is that voice and data are delivered separately to separate end devices, negating the availability of truly converged and-user applications. This solution is usually deployed as an interim solution until true, standards-based QoS is available on the wireless LAN.

802.11e

The 802.11e task group is currently refining the 802.11 MAC (Medium Access Layer) to improve QoS for better support of audio and video (such as MPEG-2) applications. The 802.11e group should finalize the standard by the end of 2002, with products probably available by mid-2003.

Because 802.11e falls within the MAC Layer, it will be common to all 802.11 versions (a,b,g, etc.) and be backward compatible with existing 802.11 wireless

LANs. As a result, the lack of 802.11e being in place today doesn't impact your decision on which version to use. In addition, upgrades should exist for your existing 802.11 access points to comply with 802.11e through relatively simple firmware upgrades once they are available.

Wireless Security

One of the most well-publicized issues with wireless LAN is security. Because wireless LANs broadcast the data in the electromagnetic spectrum, it is relatively easy to intercept the transmissions and decode them. The security of a wireless LAN is very important, especially for applications hosting valuable information. For example, networks transmitting credit card numbers for verification or storing sensitive information are definitely candidates for emphasizing security. By placing voice calls over the WLAN, vulnerability is increased, as those who intercept and decode the wireless packets could then listen in on conversations. Because of this vulnerability there have been several security mechanisms developed for the wireless LAN.

WEP

WEP (wired equivalent privacy) is 802.11's optional encryption standard implemented in the MAC Layer that most radio network interface card (NIC) and access point vendors support.

If a user activates WEP, the NIC encrypts the payload (frame body and CRC) of each 802.11 frame before transmission, using an RC4 stream cipher provided by RSA Security. The receiving station, such as an access point or another radio NIC, performs decryption upon arrival of the frame. As a result, 802.11 WEP only encrypts data between 802.11 stations. Once the frame enters the wired side of the network, such as between access points, WEP no longer applies.

As part of the encryption process, WEP prepares a keyschedule ("seed") by concatenating the shared secret key supplied by the user of the sending station with a random-generated 24-bit initialization vector (IV). The IV lengthens the life of the secret key because the station can change the IV for each frame transmission. WEP inputs the resulting "seed" into a pseudo-random number generator that produces a keystream equal to the length of the frame's payload plus a 32-bit integrity check value (ICV).

The ICV is a check sum that the receiving station eventually recalculates and compares to the one sent by the sending station to determine whether the transmitted data underwent any form of tampering while in transit. If the receiving station calculates an ICV that doesn't match the one found in the frame, then the receiving station can reject the frame or flag the user.

WEP specifies a shared secret 40- or 64-bit key to encrypt and decrypt the data. Some vendors also include 128-bit keys (know as "WEP2") in their products. With WEP, the receiving station must use the same key for decryption. Each radio NIC and access point, therefore, must be manually configured with the same key.

Before transmission takes place, WEP combines the keystream with the payload/ICV through a bitwise XOR process, which produces ciphertext (encrypted data). WEP includes the IV in the clear (unencrypted) within the first few bytes of the frame body. The receiving station uses this IV along with the shared secret key supplied by the user of the receiving station to decrypt the payload portion of the frame body.

In most cases, the sending station will use a different IV for each frame (this is not required by the 802.11 standard). When transmitting messages have a common beginning, such as the "FROM" address in an e-mail, the beginning of each encrypted payload will be equivalent when using the same key. After encrypting the data, the beginnings of these frames would be the same, offering a pattern that can aid hackers in cracking the encryption algorithm. Because the IV is different for most frames, WEP guards against this type of attack. The frequent changing of IVs also improves the ability of WEP to safeguard against someone compromising the data.

WEP has been part of the 802.11 standard since initial ratification in September 1999. At that time, the 802.11 committee was aware of some WEP limitations; however, WEP was the best choice to ensure efficient implementations worldwide. Nevertheless, WEP has undergone much scrutiny and criticism over the past couple of years.

WEP is vulnerable because of relatively short IVs and keys that remain static. The issues with WEP don't really have much to do with the RC4 encryption algorithm. With only 24 bits, WEP eventually uses the same IV for different data packets. For a large busy network, this reoccurrence of IVs can happen within an hour or so. This results in the transmission of frames having keystreams that are too similar. If a hacker collects enough frames based on the same IV, the individual can determine the shared values among them, i.e., the keystream or the shared secret key. This of course leads to the hacker decrypting any of the 802.11 frames.

The static nature of the shared secret keys emphasizes this problem. 802.11 doesn't provide any functions that support the exchange of keys among stations. As a result, system administrators and users generally use the same keys for weeks, months, and even years. This gives mischievous culprits plenty of time to monitor and hack into WEP-enabled networks. Some vendors deploy dynamic key distribution solutions based on 802.1X, which improves the security of wireless LANs. The problem, however, is that these types of mechanisms won't be part of the 802.11 standard until the end of 2003 at best.

Despite the flaws, WEP is better than nothing, and WEP is often used as a minimum level of security. Many people have taken to the streets to discover wireless LANs in neighborhoods, business areas, and colleges using protocol analyzers, such as *AiroPeek* and *Airmagnet*. Most of these people are capable of detecting wireless LANs where WEP is not in use and then using a laptop to gain access to resources located on the associated network.

There are proprietary enhancements to WEP that leading wireless LAN vendors currently implement (such as Agere's 152-bit WEP and US Robotic's 256-bit WEP), and some companies utilize Internet-based security mechanisms (e.g., IPSec) to protect data transmissions from eavesdroppers. For a standardized solution, the 802.11i committee is nearly finished specifying methods that strongly enhance 802.11's ability to safeguard wireless LANs.

802.1X

Combined with an authentication protocol, such as EAP-TLS, LEAP, or EAP-TTLS, IEEE 802.1X provides port-based access control and mutual authentication between clients and access points via an authentication server. The use of digital certificates makes this process very effective. 802.1X also provides a method for distributing encryption keys dynamically to wireless LAN devices, which solves the key reuse problem found in the current version of 802.11.

The use of IEEE 802.1X offers an effective framework for authenticating and controlling user traffic to a protected network, as well as dynamically varying encryption keys. 802.1X ties a protocol called EAP (Extensible Authentication Protocol) to both the wired and wireless LAN media and supports multiple authentication methods, such as token cards, Kerberos, one-time passwords, certificates, and public key authentication.

Initial 802.1X communications begin with an unauthenticated supplicant (i.e., client device) attempting to connect with an authenticator (i.e., 802.11 access point). The access point responds by enabling a port for passing only EAP packets from the client to an authentication server located on the wired side of the access point. The access point blocks all other traffic, such as HTTP, DHCP, and POP3 packets, until the access point can verify the client's identity using an authentication server (e.g., RADIUS). Once authenticated, the access point opens the client's port for other types of traffic.

The basic 802.1X protocol provides effective authentication regardless of whether 802.11 WEP keys or no encryption at all are implemented. Most major wireless LAN vendors, however, are offering proprietary versions of dynamic key management using 802.1X as a delivery mechanism. If configured to implement dynamic key exchange, the 802.1X authentication server can return session keys to the access point along with the accept message. The access point uses the session keys to build, sign, and encrypt an EAP key message that is sent to the client immediately after sending the success message. The client can then use the contents of the key message to define applicable encryption keys. In typical 802.1X implementations, the client can automatically change encryption keys as often as necessary to minimize the possibility of eavesdroppers having enough time to crack the key in current use.

Microsoft supports 802.1X in Windows XP, and many vendors offer 802.1X in wireless LAN devices. 802.11i is including 802.1X in the future 802.11 standard, which will be available by 2003.

TKIP

The temporal key integrity protocol (TKIP), initially referred to as WEP2, is an interim solution that fixes the key reuse problem of WEP, that is, periodically using the same key to encrypt data. The TKIP process begins with a 128-bit "temporal key" shared among clients and access points. TKIP combines the temporal key with the client's MAC address and then adds a relatively large 16-octet initialization vector to produce the key that will encrypt the data. This procedure ensures that each station uses different key streams to encrypt the data.

TKIP uses RC4 to perform the encryption, which is the same as WEP. A major difference from WEP, however, is that TKIP changes temporal keys every 10,000 packets. This provides a dynamic distribution method that significantly enhances the security of the network.

An advantage of using TKIP is that companies having existing WEP-based access points and radio NICs can upgrade to TKIP through relatively simple firmware patches. In addition, WEP-only equipment will still interoperate with TKIP-enabled devices using WEP. TKIP is a temporary solution, and most experts believe that stronger encryption is still needed.

AES

In addition to the TKIP solution, the 802.11i standard will likely include the Advanced Encryption Standard (AES) protocol. AES offers much stronger encryption. In fact, the U.S. Commerce Department's National Institutes of Standards and Technology (NIST) organization chose AES to replace Data Encryption Standard (DES). AES is now a Federal Information Processing Standard, FIPS Publication 197, that defines a cryptographic algorithm for use by U.S. government organizations to protect sensitive, unclassified information.

An issue, however, is that AES requires a coprocessor (additional hardware) to operate. This means that companies need to replace existing access points and client NICs to implement AES. Based on marketing reports, the installed base today is relatively small compared to what future deployments will bring. As a result, there will be a very large percentage of new wireless LAN implementations that will readily take advantage of AES when it becomes part of 802.11.

Companies having installed wireless LANs, on the other hand, will need to determine whether it's worth the costs of upgrade for better security.

WLAN Security Guidelines

The following guidelines are ways to provide minimum-level security in a wireless LAN. These techniques can help guard against casual network "snoopers" and provide adequate security for less-sensitive applications.

Disable SSID Broadcasting

This ensures that the access point doesn't include the SSID (service set identifier) in the beacon frames that are sent multiple times per second. Without the broadcasting of SSIDs, operating systems such as Windows XP will not discover the SSID and automatically configure the user's radio NIC. As a result, an intruder will have to find out the SSID through other, more difficult means. 802.11 association frames always include the SSID, even when SSID broadcasting is off. Thus, someone can use an 802.11 packet analyzer (and sniff the air while a legitimate user boots up and associates with an access point. This requires enough effort (and expense) to cause most casual intruders to go elsewhere. In some cases, though, it may not be practical to turn off SSID broadcasting. For example, you should broadcast SSIDs in public wireless LANs to provide open connectivity.

Utilize Static IP Addresses

By default, most wireless LANs utilize DHCP (dynamic host configuration protocol) to more efficiently assign IP addresses automatically to user devices. With a proper SSID, anyone implementing DHCP will obtain an IP address automatically and become a genuine node on the network. By disabling DHCP and assigning static IP addresses to all wireless users, you can minimize the possibility of the hacker obtaining a valid IP address. This limits their ability to access network services. The use of static IP addresses in larger networks can be very cumbersome, which may prompt network managers to use DHCP to avoid support issues.

Enable WEP

Even with all of the problems listed with WEP, it still provides better security than no security and can deter more casual intruders.

Utilize Shared Key Authentication

Most wireless LANs on the market today allow the use of this optional 802.11 feature. When the authentication process occurs, the access point sends the radio NIC a string of challenge text. The radio NIC must encrypt the challenge text with its WEP key and send the encrypted version to the access point. After decrypting the challenge text with the common WEP key, the access point can determine that the radio NIC has the correct key if the challenge text matches what was sent initially. This forms the basis for allowing the NIC to authenticate with the access point.

Use Personal Firewalls

Keep files in access-protected directories to avoid others from stealing them. Of course this applies to wired networks as well.

In more sensitive environments, more security methods can be deployed to further increase the security on the wireless LAN.

Virtual Private Networks (VPN)

This involves the use of third-party encryption (e.g., triple Data Encryption Standard or 3DES) that affects all data on the WLAN. Generally, the user installs VPN client software on their wireless device, which communicates securely with the VPN network. This can be a relatively expensive and somewhat inflexible solution, but it provides excellent security.

Mutual Authentication Mechanisms

The addition of a RADIUS server, 802.1X protocols, and possibly an access controller, provides a framework for deploying mutual authentication between users and access points. This reduces man-in-the-middle attacks, such as rogue access points. Many enterprise-grade access points support these features. 802.1X provides port-based access control and mutual authentication between clients and access points via an authentication server, such as RADIUS. Multiple authentication types are available, such as EAP-TLS or EAP-TTLS. Encryption of user names and passwords or digital certificates strengthens the authentication process. 802.1X also provides a method for distributing encryption keys dynamically to wireless LAN devices, which solves the key reuse problem found in the current version of 802.11 WEP.

Firewalls

In order to provide a higher level of security, the wireless LAN access points can be placed outside the firewall. The firewall can then be configured to enable access from legitimate users based on MAC addresses. Many commercially available access points have a MAC address filtering mechanism built in.

Minimize Excess Coverage

By properly orienting antennas to avoid covering areas outside the physically controlled boundaries of the facility, the ability for an intruder to participate on the wireless LAN is significantly reduced. This will also minimize the impact of someone disabling the wireless LAN with jamming techniques.

Public Wireless

The demand for converged voice and data services is not only growing in the on-campus wireless networks, but also on public wireless networks. As converged voice/data/video user applications such as unified messaging, presence services, intelligent agents, and multi-service contact centers grow in the on-premise environment, the demand for access to these services outside of the corporate office will grow. This anticipated demand, plus the general public's desire for better and more useful wireless services, is already prompting large expansions of public converged networks. These public networks are typically deployed on one of two technologies—public WiFi access or next-generation cellular services, commonly called third generation or 3G services.

Public WiFi Networks

Many public locations are beginning to deploy 802.11-based wireless LANs (often called WiFi networks) for the use of their customers. These deployments are becoming more and more common in places such as college campuses, airports, coffee shops, and many other public areas. In the WiFi community these are often called "HotSpots."

Public 802.11 networks can take advantage of all of the features for services and security discussed in the previous section plus features like RADIUS and 802.1x can be used for user authentication and to build billing applications. All of these capabilities can be used along with remote access schemes to give users access to their converged network services from these public networks. This greatly extends the reach of converged applications to remote workers, home users, and mobile users.

Because of the popularity of 802.11-based wireless networking and the large number of available 802.11b, and soon 802.11a, endpoints, public carriers and their associated device manufacturers are beginning to develop public services directly based on the WiFi standards. There are even developments underway to deploy 802.11-based cellular telephones and converged voice/data devices.

Next-Generation Wireless Networks (3G)

Existing second-generation public wireless networks (such as GSM) have only limited data capability (typically around 9.6 Kbps). This means that these services have not proved terribly satisfactory to support true enterprise-level converged applications. The advent of third-generation (3G) wireless communications promises a truly mobile data network with real-time Internet connectivity.

The concept of 3G wireless technologies is a major shift from circuit-switched, voice-only services, to services that are packet-based and multimedia-oriented (voice, data, video, fax). The third generation of mobile communications will greatly enhance the implementation of sophisticated wireless applications. Users will be able to utilize personal, location-based wireless information and interactive services. Also, many companies and corporations are looking into restructuring their business processes to be able to fully exploit the opportunities provided by the emerging new wireless data services.

3G is considered the future because the vastly increased bandwidth will eventually allow data transmission speeds many times faster than with a traditional GSM connection. At these new speeds, large amounts of data can be transmitted every second, the sort of data volumes that make it possible to send and receive pictures and video images and download music files using mobile devices. This is achieved by using packet-switching (instead of circuit-switching) techniques, bringing mobile communication into the world of Internet Protocol (IP). This can open up the potential to turn the mobile device in your pocket into a personal resource center—not only for communicating, but for entertainment, for referencing information, for conducting m-commerce (e-commerce for mobiles), and for managing your business affairs.

3G Network Architecture

3G will require operators to establish new Base Station Systems (BSS) with specialized Radio Network Controllers. New cell layouts, with a denser deployment of base stations, will be required to handle the new 3G wavelengths. Special media gateways will also be needed to manage all the protocol conversion processes inherent in a multimedia environment. The 3G backbone network will need to support packet-switching techniques such as ATM and IP.

Planned Applications

The most commonly touted applications to be supported by 3G networks are

- *Pictures*—Transmitting photos, designs, graphic images.
- *Video*—3G will be able to display and transmit high-quality video images. Devices with built-in video cameras will also be able to generate and

transmit moving images in real time. Video clips will also be able to be captured from standard camcorders that feature Bluetooth.

- *Audio*—Sound and music files can be downloaded and transferred using technologies such as MP3, which may be built into some devices. Sound files may also be streamed over 3 G.
- *Voice*—Voice over IP (VoIP) allows telephone calls to be routed over the Internet, either to another mobile *device* or to an endpoint on a converged IP network.
- *Software downloads*—3 G will provide a convenient means of purchasing software for instant delivery over the network. It also offers opportunities for ASPs to deliver hosted applications charged on a usage basis.
- *Roaming*—3 G's Virtual Home Environment (VHE) provides roaming users with a "home-from-home"—a personalized on-screen mobile environment that looks and feels the same wherever in the world they are.
- *Agents*—The concept of e-agents, commissioned by the user to search the Web for specifically defined information, is becoming a familiar part of fixed network Internet life. It is likely to become very much a standard application in 3 G where the "always-on" concept will lend itself to providing so-called "push" alerts when an agent finds what you are looking for.

Because of the fundamental change in public wireless networks needed to support 3 G services, the rollout of 3 G wireless networks across the world will be gradual. There are existing technologies that enhance the current 2 G network and which allow better data services to be implemented.

- General Packet Radio Service (GPRS) offers a technology that supplements today's Circuit Switched Data and Short Message Service on 2 G networks. GPRS involves overlaying a packet-based wireless interface onto the existing circuit-switched network, thus improving data rates up to 115 Kbps and enabling more data-intensive services.
- Enhanced Data for GSM Evolution (EDGE) represents the final evolution of data communications within the GSM standard. EDGE uses a modulation scheme that enables data throughput speeds of up to 384 Kbps using an existing GSM infrastructure. EDGE offers incumbent GSM operators that do not have third-generation licenses an alternative route to providing data services.

Public Wireless Directions

There is currently a great debate in the public wireless services market as to how these next-generation networks will roll out.

WiFi offers the use of already existing and relatively stable hardware and endpoints. The major drawback for WiFi is that it was not originally designed as a public service infrastructure, so there are issues with user registration and billing that make it difficult for carriers to deploy as a charged, public service.

3 G is designed around the public carrier infrastructure with customer support and billing systems included, but will take the development of new network infrastructure and endpoints.

Many industry experts see the final public network as being a mixture of these two technologies with WiFi in "HotSpot" areas and campuses and 3 g in more open environments such as public cellular networks.

However the final network is implemented, the roll-out of these types of public services has the potential to cause a vast increase in the demand for, and usage of, converged services and devices. In the long run, this may be the main driver of public acceptance of the new, converged applications for business.

A Quick Overview of Mobile IP

One of the technologies being developed to help support seamless converged IP voice, data, and video applications into a mobile environment is the Mobile IP concept. Mobile IP is a technology that allows IP applications to move from one network to another. An example of where this might be used is for allowing an office worker to use their converged IP voice/data PDA both in the office and on a public network, when out of the office.

IP routes packets from a source endpoint to a destination by allowing routers to forward packets from incoming network interfaces to outbound interfaces according to routing tables. The routing tables typically maintain the next-hop (outbound interface) information for each destination IP address, according to the number of networks to which that IP address is connected. The network number is derived from the IP address by masking off some of the low-order bits. Thus, the IP address typically carries with it information that specifies the IP node's point of attachment. To maintain existing transport-layer connections as the mobile node moves from place to place, it must keep its IP address the same. In TCP (which accounts for the overwhelming majority of Internet connections), connections are indexed by a quadruplet that contains the IP addresses and port numbers of both connection endpoints. Changing any of these four numbers will cause the connection to be disrupted and lost. On the other hand, correct delivery of packets to the mobile node's current point of attachment depends on the network number contained within the mobile node's IP address, which changes at new points of attachment. To change the routing requires a new IP address associated with the new point of attachment.

Mobile IP (IETF RFC 2002) has been designed to solve this problem by allowing the *mobile node* to use two IP addresses. In Mobile IP, the *home address* is static and is used, for instance, to identify TCP connections. The *care-of address* changes at each new point of attachment and can be thought of as the mobile node's topologically significant address; it indicates the network number and thus identifies the mobile node's point of attachment with respect to the network topology. The home address makes it appear that the mobile node is continually able to receive data on its *home network*, where Mobile IP requires the existence of a network node known as the *home agent*.

Whenever the mobile node is not attached to its home network (and is therefore attached to what is termed a *foreign network*), the home agent gets all the packets destined for the mobile node and arranges to deliver them to the mobile node's current point of attachment.

Whenever the mobile node moves, it *registers* its new care-of address with its home agent. To get a packet to a mobile node from its home network, the home agent delivers the packet from the home network to the care-of address. The further delivery requires that the packet be modified so that the care-of address appears as the destination IP address. This modification can be understood as a packet transformation or, more specifically, a *redirection*. When the packet arrives at the care-of address, the reverse transformation is applied so that the packet once again appears to have the mobile node's home address as the destination IP address. When the packet arrives at the mobile node, addressed to the home address, it will be processed properly by TCP or whatever higher-level protocol logically receives it from the mobile node's IP (that is, layer 3) processing layer.

In Mobile IP, the home agent redirects packets from the home network to the care-of address by constructing a new IP header that contains the mobile node's care-of address as the destination IP address. This new header then shields or encapsulates the original packet, causing the mobile node's home address to have no effect on the encapsulated packet's routing until it arrives at the care-of address. Such *encapsulation* is also called *tunneling*, which suggests that the packet burrows through the Internet, bypassing the usual effects of IP routing.

Some technical problems with Mobile IP remain. The largest one is security in a public and private network. However, once the security solutions are solid, nomadic users may finally begin to enjoy the convenience of seamless roaming and effective application transparency that is the promise of Mobile IP.

Summary

Wireless networking is one of the fastest growing areas in the entire IT industry. As wireless networks proliferate both in private enterprises and in public networks, more and more users will be utilizing both voice and data communications across this network.

In this chapter we have discussed the fundamental wireless LAN components and protocols and how they can support both voice and data traffic. We have also discussed public wireless access methods such as Wi-Fi "hotspots" as well as emerging "3 G" applications.

As users see the benefits of accessing both data and voice wirelessly, there will be an ever-increasing demand for support of converged applications on a wireless infrastructure.

CHAPTER 9

CONVERGENCE STANDARDS AND PROTOCOLS

The convergence of voice and data networks has begun to drive radical changes in the development and delivery of products for the service provider, whether a small, medium-sized, or large business enterprise. The absolute goal of these developments is to someday create a single network fabric over which both voice and data are delivered seamlessly and effectively to any type of communication device. Today we are at the beginning stages of this drive toward convergence. It is important to consider the ways in which convergence products and communication standards and protocols will evolve into solutions that provide meaningful business advantages. Customers, business partners, and manufacturers are all talking about convergence of networks. The drive for the convergence networks is spurred by the unprecedented growth in one particular network type, the IP-based Internet and its innovative service offering.

The growth in Internet traffic is an unrelenting force that is opening a vast array of new information and transaction opportunities, both for end users using public tele- and data communication networks, and for enterprises using services that optimize and streamline their businesses. Common to both is the concept of a network that supports whatever content and whatever applications are necessary to satisfy their needs.

As we all know, IP was originally conceived as a transport protocol for data traffic, not real-time voice and video traffic. As a result, when IP networks are heavily loaded, real-time delivery of voice is affected by serious quality issues due to the lack of provisioning for voice traffic. These issues include delay, jitter, and packet loss. So maintaining QS is a major part of the converged network. Because standards for IP voice transmission have not yet matured, interoperability issues exist at all points in the IP network. Today, mainly proprietary voice compression algorithms are used to packetize voice. This means that only like devices can communicate over an IP network. Standards such as Session Initiated Protocol (SIP) and H.323 are maturing to support convergence and next-generation solutions. To understand convergence it is important to explore the standards and protocols that support convergence technologies.

H.323

An H.323 gateway is an endpoint on the network that provides for real-time, two-way communications between H.323 terminals on the IP network and other ITU terminals on a switch-based network, or to another H.323 gateway. It performs the function of a "translator," performing the translation between different transmission formats (e.g., from H.225 to H.221). It is also capable of translating between audio and video codecs. The gateway is the interface between the PSTN and the Internet. It takes voice from circuit-switched PSTN and places it on the public Internet and vice versa.

Gateways are optional in that terminals in a single LAN can communicate with each other directly. When the terminals on a network need to communicate with an endpoint in some other network, then they communicate via gateways using the H.245 and Q.931 protocols. H.323 is the ITU-T's (International Telecommunications Union) standard with which vendors should comply while providing Voice over IP service. This recommendation provides the technical requirements for voice communication over LANs while assuming that no Quality of Service (QoS) is being provided by LANs. It was originally developed for multimedia conferencing on LANs, but was later extended to cover Voice over IP. The first version was released in 1996 while the second version of H.323 came into effect in January 1998. The standard encompasses both point-to-point communications and multipoint conferences. The products and applications of different vendors can interoperate if they abide by the H.323 specification. H.323 defines four logical components: Terminals, Gateways, Gatekeepers, and Multipoint Control Units (MCUs). Terminals, gateways, and MCUs are known as endpoints. These are discussed next.

Terminals

These are the LAN client endpoints that provide real-time, two-way communications. All H.323 terminals have to support H.245, Q.931, Registration Admission Status (RAS), and Real Time Transport Protocol (RTP). H.245 is used for allowing the usage of the channels, Q.931 is required for call signaling and setting up the call, RTP is the real time transport protocol that carries voice packets, while RAS is used for interacting with the gatekeeper. H.323 terminals may also include T.120 data conferencing protocols, video codecs, and support for MCU. An H.323 terminal can communicate with either another H.323 terminal, a H.323 gateway, or a MCU.

Gatekeepers

A gatekeeper is the most vital component of the H.323 system, dispatching the duties of a "manager." It acts as the central point for all calls within its zone (a zone is the aggregation of the gatekeeper and the endpoints registered with it) and provides services to the registered endpoints. Some of the functionalities that gatekeepers provide are listed following:

Address Translation: Translation of an alias address, which provides an alternate method of addressing the endpoint. It could be an email address, a telephone number, or something similar. An endpoint may have one or more alias addresses associated with it and each alias is unique within a zone to a transport address. This is done using a translation table that is updated by using the registration messages.

Admissions Control: Gatekeepers can either grant or deny access based on call authorization, source and destination addresses, or some other criteria.

Call signaling: The Gatekeeper may choose to complete the call signaling with the endpoints and may process the call signaling itself. Alternatively, the Gatekeeper may direct the endpoints to connect the Call Signaling Channels directly to each other.

Call Authorization: The Gatekeeper may reject calls from a terminal due to authorization failure through the use of H.225 signaling. The reasons for rejection could be restricted access during some time periods or restricted access to/from particular terminals or Gateways.

Bandwidth Management: The Gatekeeper controls the number of H.323 terminals permitted simultaneous access to the network. Through the use of H.225 signaling, the Gatekeeper may reject calls from a terminal due to bandwidth limitations.

Call Management: The gatekeeper may maintain a list of ongoing H.323 calls. This information may be necessary to indicate that a called terminal is busy, and to provide information for the Bandwidth Management function.

Multipoint Control Units (MCU)

The MCU is an endpoint on the network that provides the capability for three or more terminals and gateways to participate in a multipoint conference. The MCU consists of a mandatory Multipoint Controller (MC) and optional Multipoint Processors (MP). The MC determines the common capabilities of the terminals by using H.245 but it does not perform the multiplexing of audio, video, and data. The multiplexing of media streams is handled by the MP under the control of the MC. Figure 9.1 is a sample of H.323 components.

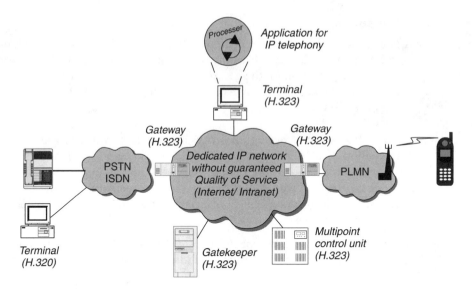

Figure 9.1

323 componenets.

H.323 Protocol Stack

Figure 9.2 shows the H.323 protocol stack. The audio, video, and registration packets use the unreliable User Datagram Protocol (UDP), while the data and control application packets use the reliable Transmission Control Protocol (TCP) as the transport protocol. The T.120 protocol is used for defining the data conferencing part. The H.323 stack is part of a continually evolving family of WAN software building blocks that provides a valuable head start to developers of telecom equipment. The stack is a result of comprehensive experience in development of telecom products for customers worldwide.

Figure 9.2

H.323 protocols in relation to the OSI model.

Data	Control and signaling		Audio/ video	Registration
T.120	H.225.0 call signaling	H.245 conference control	RTP/RTCP	H.225.0 RAS
TCP			UDP	
Network layer				
Data link layer				
Physical layer				

Let's discuss the H.323 protocol stack in detail. Control and Signaling in H.323 provide three control protocols: H.225.0/Q.931 Call Signaling, H.225.0 RAS, and H.245 Media Control. H.225/ Q.931 is used in conjunction with H.323 and provides the signaling for call control. For establishing a call from a source to a receiver host, the H.225 RAS (Registration, Admission, and Signaling) channel is used. After the call has been established, H.245 is used to negotiate the media streams.

H.225.0: RAS

The RAS channel is used for communication between the endpoints and the gatekeeper. Because the RAS messages are sent over UDP (an unreliable channel), it recommends timeouts and retry counts for messages. The procedures defined by the RAS channel are H.323.

Gatekeeper Discovery

This is the process that an endpoint uses to determine the gatekeeper with which it should register. The endpoint normally multicasts a Gatekeeper Request (GRQ) message asking for its gatekeeper. One or more gatekeepers may respond with the Gatekeeper Confirmation (GCF) message, thereby indicating willingness to be the gatekeeper for that endpoint. The response includes the transport address of the gatekeeper's RAS channel. Gatekeepers who do not want the endpoint to register with them can send a Gatekeeper Reject (GRJ) message. If more than one gatekeeper responds with GCF, then the endpoint may choose the gatekeeper and register with it. If no gatekeeper responds within a timeout interval, the endpoint may retransmit the GRQ.

Endpoint Registration

This is the process by which an endpoint joins a zone and informs the gatekeeper of its transport and alias addresses. All endpoints usually register with the gatekeeper that was identified through the discovery process. An endpoint shall send a Registration Request (RRQ) to a gatekeeper. This is sent to the gatekeeper's RAS channel Transport Address. The endpoint has the network address of the gatekeeper from the gatekeeper discovery process and uses the well-known RAS channel TSAP Identifier. The gatekeeper shall respond with either a Registration Confirmation (RCF) or a Registration Reject (RRJ). The gatekeeper shall ensure that each alias address translates uniquely to a single transport address. An endpoint may cancel its registration by sending an Unregister Request (URQ) message to the gatekeeper. The gatekeeper shall respond with an Unregister Confirmation (UCF) message. A gatekeeper may cancel the registration of an endpoint by sending an Unregister Request (URQ) message to the endpoint. The endpoint shall respond with an Unregister Confirmation (UCF) message.

Endpoint Location

An endpoint or gatekeeper that has an alias address for an endpoint and would like to determine its contact information may issue a Location Request (LRQ) message. The gatekeeper with whom the requested endpoint is registered shall respond with the Location Confirmation (LCF) message containing the contact information of the endpoint or the endpoint's gatekeeper. All gatekeepers with whom the requested endpoint is not registered shall return Location Reject (LRJ) if they received the LRQ on the RAS channel.

Admissions, Bandwidth Change, Status, and Disengage

The RAS channel is also used for the transmission of Admissions, Bandwidth Change, Status, and Disengage messages. These messages are exchanged between an endpoint and a gatekeeper and are used to provide admissions control and bandwidth management functions. The Admissions Request (ARQ) message specifies the requested call bandwidth. The gatekeeper may reduce the requested call bandwidth in the Admissions Confirm (ACF) message. An endpoint or the gatekeeper may attempt to modify the call bandwidth during a call using the Bandwidth Change Request (BRQ) message.

H.225.0 Call Signaling

The call signaling channel is used to carry H.225 control messages. In networks that do not contain a gatekeeper, call signaling messages are passed directly between the calling and called endpoints using the Call Signaling Transport Addresses. It is assumed that the calling endpoint knows the Call Signaling Transport Address of the called endpoint and thus can communicate directly. In networks that do contain the gatekeeper, the initial admission message exchange takes place between the calling endpoint and the gatekeeper using the gatekeeper's RAS channel transport address. The call signaling is done over TCP (reliable channel).

Call Signaling Channel Routing

Call Signaling messages may be passed in two ways. The first way is Gatekeeper Routed Call Signaling, where the call signaling messages are routed through the gatekeeper between the endpoints. The other alternative is Direct Endpoint Call Signaling, where the call signaling messages are passed directly between the endpoints. Admissions messages are exchanged with the gatekeeper over the RAS channel, followed by an exchange of call signaling messages on a Call Signaling Channel, which in turn is followed by the establishment of the H.245, Control Channel.

Control Channel Routing

When gatekeeper routed call signaling is used, there are two methods to route the H.245 Control Channel. The first alternative establishes the H.245 Control Channel directly between the endpoints, while in the second case, the establishment of the H.245 Control Channel is done through the gatekeeper.

H.245 Media and Conference Control

H.245 is the media control protocol that H.323 systems utilize after the call establishment phase has been completed. H.245 is used to negotiate and establish all of the media channels carried by RTP/RTCP. The functions offered by H.245 are

- Determining master and slave: H.245 appoints a Multipoint Controller (MC) that is held responsible for central control in cases where a call is extended to a conference.
- Capability Exchange: H.245 is used to negotiate the capabilities when a call has been established. The capability exchange can occur at any time during a call, thereby allowing renegotiations at any time.
- Media Channel Control: After conference endpoints have exchanged capabilities, they may open and close logical channels of media. Within H.245, media channels are abstracted as logical channels (which are just identifiers).
- Conference Control: In conferences, H.245 provides the endpoints with mutual awareness and establishes the media flow model between all the endpoints.

Call Setup in H.323

The procedure to set up a call involves

- Discovering a gatekeeper that would take the management of that endpoint
- Registration of the endpoint with its gatekeeper
- Endpoint entering the call setup phase
- The capability exchange taking place between the endpoint and the gatekeeper
- The call being established
- When the endpoint is done, terminating the call. The termination can also be initiated by the gatekeeper.

H.323 is a derivative of the H.320 videoconferencing standard, but it assumes LAN connectivity rather than ISDN between conferencing components. As such, QoS is not assumed and is not implicitly supported. When used to support a VoIP application, the calls are treated as audio-only videoconferences. Standards-based videoconferencing is generally governed by the International Telecommunications Union (ITU) "H-series" recommendations, which include H.320 (ISDN protocol), H.323 (LAN protocol), and H.324 (POTS protocol). These standards specify the manner in which real-time audio, video, and data communications take place over various communications topologies. Standards compliance promotes common capabilities and interoperability between networked multimedia building blocks that may be provided by multiple vendors. The H.323 standard was ratified in 1996 and consists of the following component standards:

- **H.225**—Specifies messages for call control, including signaling, registration, and admissions, and packetization/synchronization of media streams.
- **H.245**—Specifies messages for opening and closing channels for media streams and other commands, requests, and indications.
- **H.261**—Video codec for audiovisual services at P ¥ 64 Kbps.
- **H.263**—Specifies a new video codec for video POTS.
- **G.711**—Audio codec, 3.1 kHz at 48, 56, and 64 Kbps (normal telephony).
- **G.722**—Audio codec, 7 kHz at 48, 56, and 64 Kbps; ratified.
- **G.728**—Audio codec, 3.1 kHz at 16 Kbps.
- **G.723**—Audio codec, for 5.3 and 6.3 Kbps modes.
- **G.729**—Audio codec (G.729a is a reduced complexity variant).

H.323 is designed to run on common network architectures. As network technology evolves and bandwidth management techniques improve, H.323-based solutions will be able to take advantage of those enhanced capabilities. The following matrix gives you an idea of H.323 network independence.

Approval Date	**H.320** **1990**	**H.321** **1995**	**H.322** **1995**	**H.323** **V1/V2** **1996/1998**	**H.324** **1996**
Network	Narrowband switched digital ISDN	Broadband ISDN ATM LAN	Guaranteed bandwidth packet-switched networks	Non-guaranteed bandwidth packet-switched networks, (Ethernet)	PSTN or POTS, the analog phone system
Video	H.261 H.263	H.261 H.263	H.261 H.263	H.261 H.263	H.261 H.263
Audio	G.711 G.722 G.728	G.711 G.722 G.728	G.711 G.722 G.728	G.711 G.722 G.728 G.723 G.729	G.723

Matrix 9.1

Comparision of H.32x family

Approval Date	H.320 1990	H.321 1995	H.322 1995	H.323 V1/V2 1996/1998	H.324 1996
Multiplexing	H.221	H.221	H.221	H.225.0	H.223
Control	H.230	H.242	H.242	H.245	H.245
	H.242		H.230		
Multipoint	H.231	H.231	H.231	H.323	
	H.243	H.243	H.243		
Data	T.120	T.120	T.120	T.120	T.120
Comm.	I.400	AAL	I.400&	TCP/IP	V.34 Modem
Interface		I.363	TCP/IP		
		AJM			
		I.361			
		PHY			
		I.400			

Matrix 9.1

Continued

The H.323 recommendation for multimedia conferencing was first ratified in late 1996. It therefore has a head start and is also widely supported by vendors. However, it has not made any significant inroads into the traditional telephony market. H.323 is a large, complex, and flexible standard and it faces some challenges. Different interpretations and oversight in implementations have led to non-compliant and incompatible devices. This issue is significant and considered to be the primary impediment to the wider deployment of H.323-enabled products and services in the enterprise, small business, and residential markets. The complexity and the flexibility of H.323 make its implementation difficult and prone to errors and omissions. Vendors of H.323 products and services often choose to implement a subset of H.323 that meets their immediate requirements. Furthermore, ITU-T does not provide an implementation guideline that can help ensure compliance and interoperability.

All endpoints must inter-operate before an H.323 multimedia call can be established. Consider, for example, a phone-to-phone call over the Internet. This setup implies that the phones (terminals) are interoperable, the gateways are interoperable, and the phones are interoperable with their respective gateways. If the endpoints are manufactured by different vendors, there is a good possibility that problems may occur. The implementation of codecs is well-developed. The issue that breaks interoperability is the capability exchange and signaling process between endpoints, which is often not implemented completely by vendors. The poor interoperability of H.323 endpoints is widely recognized. The International Multimedia Teleconferencing Consortium (IMTC) was set up with the primary goal of ensuring that vendors' products and services are interoperable. IMTC is a non-profit organization with over 150 members, regularly conducting interoperability and compliance testing of products and services.

The vision for H.323 is global interoperability between packet- and circuit-switched networks. H.323 also promises new and integrated services of value to customers currently using circuit-switching technologies exclusively. These goals have not yet been achieved. From a business point of view, a technology cannot succeed solely on the basis of lower operational costs. It must also provide additional value and performance in the form of ease of use and improved feature set for it to supplant a well-entrenched technology such as PSTN. Internet telephony service providers (ITSP) and Internet service providers (ISP) were expected to provide this level of interoperability with better services and value as compared to plain old telephone service (POTS). Several ITSPs exist today (e.g., BizTrans) that do provide good value in certain regions such as North America and Europe. Global interoperability is still a problem. Furthermore, the features and quality of service being offered are often inferior to POTS. Two key factors have hampered ITSPs and ISPs efforts. The first is the lack of interoperability of endpoints (especially gateways) from different vendors. And the second is the poor scalability of H.323 communications. ITXC, the Internet telephony exchange carrier, has the largest global presence of H.323 gateways. As such, it is well positioned to offer services to ITSPs, ISPs, and traditional telephone companies to route their multimedia traffic over its packet-based network.

H.323 or SIP?

Now questions come regarding which protocol will win the race, H.323 or SIP. This is the question IP telephony vendors are asking. The H.323 recommendation has been around since 1996 but has failed to capture the market. The Internet Engineering Task Force (IETF) is working on parallel standards for IP telephony. The Session Initiation Protocol (SIP), an application-level protocol for establishing multimedia communications, has gained momentum. SIP proponents cite the following as advantages of SIP over H.323:

- **IP-based**: IP is the dominant protocol both at the edges and in the core of the Internet. As a result, interoperability with ATM and ISDN is not an issue. H.323 carries a lot of extra baggage to make sure that it is interoperable with the other standards in the series. SIP is free of this extra rarely needed luggage.
- **Less complex**: SIP is a much smaller and less complicated standard that is based on the architecture of existing popular protocols such as HTTP and FTP. On the other hand, H.323 is large and complicated. As a result, H.323 products and services are more expensive to develop.
- **Easy to decode/debug**: SIP uses a simple format for commands and messages. These are text strings that are easy to decode, and hence, easy to debug. The entire set of messages is also much smaller than in H.323.

- **Client/server architecture**: SIP messages are exchanged between a client and a server like HTTP messages. This client–server operation mode allows security and management features to be implemented easily in SIP when compared to H.323 calls.
- **Easier firewall/proxy design and configuration**: SIP commands can easily be proxied and firewalls can be designed to allow/disallow SIP communications. Getting H.323 through firewalls and proxies is much more complicated.
- **Extendible and scalable**: Because SIP is based on a client/server distributed architecture it is more scalable than H.323, which often requires peer-to-peer communications. Extending SIP is also easier because of its simpler message format and greater experience with similar protocols such as HTTP.

Analysts often make an analogy between H.323 and ATM as standards that provided too much too soon; the market was not ready for them. Although the share of SIP will increase, H.323 will also grow as most of its interoperability problems have been addressed. Moreover, the investment in H.323 by vendors and customers alike will prevent wholesale migration to SIP. SIP and H.323 will coexist in products and services for several years to come.

Session Initiation Protocol (SIP)

In the near future, if you make a telephone call, it's quite likely that the Session Initiation Protocol will be what finds the person you're trying to reach and causes their phone to ring. SIP is all about calling people and services. The most important way that SIP differs from the current system of making a telephone call (apart from that it's an IP-based protocol) is that you may not be dialing a number at all. Although SIP can call traditional telephone numbers, SIP's native concept of an address is a SIP URL, which looks very like an email address.

The Session Initiation Protocol (SIP) is an application-layer control protocol that can establish, modify, and terminate multimedia sessions or calls. These multimedia sessions include multimedia conferences, distance learning, Internet telephony, and similar applications. SIP can invite both persons and "robots," such as a media storage service. SIP can invite parties to both unicast and multicast sessions; the initiator does not necessarily have to be a member of the session to which it is inviting. Media and participants can be added to an existing session. SIP can be used to initiate sessions as well as invite members to sessions that have been advertised and established by other means. Sessions can be advertised using multicast protocols such as SAP, electronic mail, news groups, web pages, or directories (LDAP), among others.

SIP transparently supports name mapping and redirection services, allowing the implementation of ISDN and Intelligent Network telephony subscriber services. These facilities also enable personal mobility. In the parlance of telecommunications intelligent network services, this is defined as follows: "Personal mobility is the ability of end users to originate and receive calls and access subscribed telecommunication services on any terminal in any location, and the ability of the network to identify end users as they move. Personal mobility is based on the use of a unique personal identity (i.e., personal number)." Personal mobility complements terminal mobility, i.e., the ability to maintain communications when moving a single end system from one subnet to another. SIP supports five facets of establishing and terminating multimedia communications.

- User location: determination of the end system to be used for communication;
- User capabilities: determination of the media and media parameters to be used;
- User availability: determination of the willingness of the called party to engage in communications;
- Call setup: "ringing," establishment of call parameters at both called and calling party;
- Call handling: including transfer and termination of calls.

SIP can also initiate multi-party calls using a multipoint control unit (MCU) or fully-meshed interconnection instead of multicast. Internet telephony gateways that connect Public Switched Telephone Network (PSTN) parties can also use SIP to set up calls between them. SIP is designed as part of the overall IETF multimedia data and control architecture currently incorporating protocols such as RSVP for reserving network resources, the real-time transport protocol (RTP) for transporting real-time data and providing QOS feedback, the real-time streaming protocol (RTSP) for controlling delivery of streaming media, the session announcement protocol (SAP) for advertising multimedia sessions via multicast and the session description protocol (SDP) for describing multimedia sessions. However, the functionality and operation of SIP does not depend on any of these protocols.

SIP can also be used in conjunction with other call setup and signaling protocols. In that mode, an end system uses SIP exchanges to determine the appropriate end system address and protocol from a given address that is protocol independent. For example, SIP could be used to determine that the party can be reached via H.323, obtain the H.245 gateway and user address, and then use H.225.0 to establish the call.

In another example, SIP might be used to determine that the callee is reachable via the PSTN and indicate the phone number to be called, possibly suggesting an Internet-to-PSTN gateway to be used. SIP does not offer conference control services such as floor control or voting and does not prescribe how a conference is to be managed, but SIP can be used to introduce conference control protocols. SIP does not allocate multicast addresses. SIP can invite users to sessions with and without resource reservation. SIP does not reserve resources, but can convey to the invited system the information necessary to do this.

SIP operation works such that Callers and callees are identified by SIP addresses. When making a SIP call, a caller first needs to locate the appropriate server and send it a request. The caller can reach the callee either directly or indirectly through the redirect servers. The Call ID field in the SIP message header uniquely identifies the calls. I will briefly discuss how the protocol performs its operations.

SIP Addressing

The SIP hosts are identified by a SIP URL, which is of the form sip:password @host. A SIP address can either designate an individual or a whole group. The "objects" addressed by SIP are users at hosts, identified by a SIP URL. The SIP URL takes a form similar to a mailto or telnet URL, i.e., user@host. The user part is a user name or a telephone number. The host part is either a domain name or a numeric network address. A user's SIP address can be obtained out-of-band, can be learned via existing media agents, can be included in some mailers' message headers, or can be recorded during previous invitation interactions. In many cases, a user's SIP URL can be guessed from their email address. A SIP URL address can designate an individual (possibly located at one of several end systems), the first available person from a group of individuals, or a whole group. The form of the address, for example, sip:password@example.com, is not sufficient, in general, to determine the intent of the caller.

If a user or service chooses to be reachable at an address that is guessable from the person's name and organizational affiliation, the traditional method of ensuring privacy by having an unlisted "phone" number is compromised. However, unlike traditional telephony, SIP offers authentication and access control mechanisms and can avail itself of lower-layer security mechanisms, so that client software can reject unauthorized or undesired call attempts.

Locating a SIP Server

The client can either send the request to a SIP proxy server or it can send it directly to the IP address and port corresponding to the Uniform Request Identifier (URI).

SIP Transaction

Once the host part of the Request URI has been resolved to a SIP server, the client can send requests to that server. A request together with the responses triggered by that request make up a SIP transaction. The requests can be sent through reliable TCP or through unreliable UDP.

SIP Invitation

A successful SIP invitation consists of two requests: an INVITE followed by ACK. The INVITE request asks the callee to join a particular conference or establish a two-party conversation. After the callee has agreed to participate in the call, the caller confirms that it has received that response by sending an ACK request. The INVITE request contains a session description that provides the called party with enough information to join the session. If the callee wishes to accept the call, it responds to the invitation by returning a similar session description.

Locating a User

A callee may keep changing its position with time. These locations can be dynamically registered with the SIP server. When the SIP server is queried about the location of a callee, it returns a list of possible locations. A Location Server in the SIP system actually generates the list and passes it to the SIP server.

Changing an Existing Session

Sometimes we may need to change the parameters of an existing session. This is done by re-issuing the INVITE message by using the same Call ID but with a new body to convey the new information.

Here a basic example of a SIP operation is given where a client is inviting a participant for a call. A SIP client creates an INVITE message for arora.32@osu.edu., which is normally sent to a proxy server. This proxy server tries to obtain the IP address of the SIP server that handles requests for the requested domain. The proxy server consults a Location Server to determine this next hop server. The Location server is a non-SIP server that stores information about the next hop servers for different users. On getting the IP address of the next hop server, the proxy server forwards the INVITE to the next hop server. After the User Agent Server (UAS) has been reached, it sends a response back to the proxy server. The proxy server in turn sends back a response to the client. The client then confirms that it has received the response by sending an ACK. The exchange of messages is shown in Figure 9.3. In this case, we had assumed that the client's INVITE request was forwarded to the proxy server. However, if it had been forwarded to a redirect server then the redirect server would return the IP address of the next hop server to the client. The client then directly communicates with the UAS.

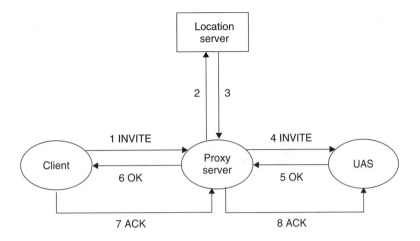

Figure 9.3

SIP operation.

Session Initiation Protocol (SIP) is a new entry into the signaling arena, with a peer-to-peer architecture much like H.323. However, unlike H.323, SIP is an Internet-type protocol in philosophy and intent. Many technologists regard SIP as a competitor to H.323 and complementary to client/server protocols such as Media Gateway Control Protocol (MGCP), that is an IP telephony signaling protocol from the IETF. MGCP was the original protocol, which evolved into MEGACO. Both protocols are designed for implementation in IP phones that are lower cost than SIP or H.323 phones. MGCP/MEGACO requires the use of softswitches for call control and more resembles the telephony model of the circuit-switched PSTN than do SIP and H.323. As such, SIP will probably see deployment in mixed environments composed of combinations of SIP endpoints along with MGCP devices.

SIP depends on relatively intelligent endpoints, which require little or no interaction with servers. Each endpoint manages its own signaling, both to the user and to other endpoints. Fundamentally, the SIP protocol provides session control, while MGCP provides device control. This provides SIP with a number of advantages. First, the simple message structure provides for call setup in fewer steps than H.323 so that performance is better than H.323 using similar processing hardware. SIP is also more scalable than H.323 because it is inherently a distributed and stateless call model. Perhaps the key difference (and advantage) of SIP is the fact that it is truly an Internet-model protocol from inception. It uses simple ASCII messaging (instead of ASN.1) based on HTTP/1.1. This means that SIP messaging is easy to decode and troubleshoot—but more importantly, it means that web-type applications can support SIP services with minimal changes. In fact, SIP fully supports URL (with DNS) naming in addition to standard E.164 North American Numbering Plan addressing. That means that in a SIP model, a user's e-mail address and phone address can be the same. It also means that the session is abstracted so that very different endpoints can communicate with each other.

SIP is modeled to support some or all of the five facets of establishing and terminating multimedia communications. Each of these facets can be discovered or negotiated in a SIP session between two endpoints.

- User location
- User capabilities
- User availability
- Call setup
- Call handling

Although SIP is philosophically a peer-to-peer protocol, it is made up of logical clients and servers, often collocated within an endpoint. For example, a typical SIP client may be an IP phone, PC, or PDA; it contains both a user agent client (UAC) to originate SIP requests and a user agent server (UAS) to terminate SIP requests. Also supported are SIP proxy servers, SIP redirect servers (RS), registrars, and location servers. These servers are all optional but also very valuable in actual SIP implementations.

SIP servers are defined here.

- **Proxy server**—Acts as a server and client; initiates SIP requests on behalf of a UAC.
- **Redirect server (RS)**—Receives a SIP request, maps the destination to one or more addresses, and responds with those addresses.
- **Registrar**—Accepts requests for the registration of a current location from UACs. Typically is collocated with a redirect server.
- **Location server**—Provides information about a callee's possible locations, typically contacted by a redirect server. A location server/service may co-exist with a SIP redirect server.

SIP is a text-based protocol with syntax much like the Hyper-Text Transfer Protocol (HTTP) and Real-Time Streaming Protocol (RTSP). A SIP *INVITE* request looks something like this:

figure = river/siprequest.ps

SIP supports several request *methods* as shown next.

Request Method	Purpose
INVITE	Invite the callee into a session
OPTIONS	Discover the capabilities of the receiver
BYE	Terminate a call or call request
CANCEL	Terminate incomplete call requests
ACK	Acknowledge a successful response
REGISTER	Register the current location of a user

In response to a request, a server sends a SIP response which gives a response code indicating how the request was processed. Response codes are divided into six categories depending on the general form of behavior expected. These categories are shown in the following table.

Response Class	Meaning	Example
1XX	Information about call status	180 RINGING
2XX	Success	200 OK
3XX	Redirection to another server	301 MOVED TEMPORARILY
4XX	Client did something wrong	401 UNAUTHORISED
5XX	Server did something wrong	500 INTERNAL SERVER ERROR
6XX	Global failure (don't resend same request)	606 NOT ACCEPTABLE

The normal progress of a SIP call involves sending an *INVITE* request, getting a *180 RINGING* response, followed by a "*200 OK*" response when the callee answers, then sending an *ACK* request to confirm the connected call.

After either party hangs up, a *BYE* request is sent, which elicits a "*200 OK*" response, which in turn causes a final *ACK* request to be sent.

The *CANCEL* method is used when a call is made to more than one destination at the same time. For example, a proxy server may respond indicating that someone might be at one of the server locations. My client can either call these one-by-one, or call all of them simultaneously. If it does the latter, then when the callee answers at one of these locations, my client sends a CANCEL request to all the other locations to stop the phone (or whatever it is) ringing there.

SIP can run over both TCP and UDP. Clearly UDP is an unreliable protocol, and so in this case SIP must provide its own reliability, which it does by retransmission. So why run SIP over UDP when TCP provides reliability for free? In truth, this reliability is not "for free." When you use TCP, you use a general-purpose reliable transport protocol, and you get TCP's concept of what reasonable retransmission timeouts are. TCP is also totally reliable—once data is put into a TCP connection, it will eventually come out the other end unless the connection is terminated. Sometimes this isn't what we want, because old data has been superseded by new data, and if the old data hasn't got there yet, we no longer want it to be delivered. In short, UDP allows the SIP applications to control the timing and reliability, so it gives improved performance to this kind of signaling protocol.

In addition, when SIP uses UDP, a proxy server may be stateless—a SIP request can be relayed without leaving behind state in the proxy because any response the request later generates contains (in its Via fields) all the information needed for the proxy to get these responses back to the caller's client application. This allows very large proxy servers to be built. If these servers used TCP, they would have to maintain TCP connection state for all calls, and this would make very large proxy servers a difficult proposition. Why then support TCP at all? The primary reason is that many firewalls do not allow UDP traffic to pass, whereas it is relatively simple to configure them to pass SIP traffic to an internal server, which may then open a hole in the firewall for the relevant RTP traffic. Eventually, though, we expect firewalls to be SIP proxies, and we envisage that SIP over TCP will not be widely used.

SIP is a very flexible, lightweight protocol for making multimedia calls. We've mostly talked about telephone-style calls because they're most familiar to most people, but SIP really comes into its own with user mobility and heterogeneous multimedia terminals. It is not yet clear whether SIP will see widespread deployment, or whether the ITU's H.323 protocol, which performs a similar task but is more heavyweight and less flexible, ends up becoming the de-facto standard. Hybrids of the two are also possible. Time and the market will decide.

ISDN Signaling—Q.931, Q.932

To deploy converged network, it is important to communicate and interoperate with the existing network and signaling infrastructure. Integrated Services Digital Network (ISDN) is essentially digital telephone service. ISDN envisions telephone companies providing "Integrated Services," capable of handling voice, data, and video over the same circuits. The core of the telephone network is now digital, so most ordinary telephone calls are now converted into bits and bytes, transported through digital circuits, and converted back into analog audio at the remote end. The international standard for the digital telephone network is Signaling System 7 (SS-7), a protocol suite in its own right, roughly comparable to TCP/IP. End users never see SS-7, because it is only used between telephone switches. ISDN provides a fully digital user interface to the SS-7 network, capable of transporting either voice or data. BISDN (Broadband ISDN) uses ATM instead of SS-7 as the underlying networking technology.

ISDN is a complete networking technology in its own right, providing clearly defined Physical, Data Link, Network, and Presentation layer protocols. For most Internet applications, though, ISDN is regarded as a fancy Data Link protocol used to transport IP packets. An ISDN interface is time division multiplexed into channels. In accordance with SS-7 convention, control and data signals are separated onto different channels. Contrast this to TCP/IP, where control packets are largely regarded as special cases of data packets and are transported over the same channel. In ISDN, the *D channel* is used for control, and the *B channels* are for data. B channels are always bi-directional 64 Kbps, the standard data rate for transporting a single audio conversation; D channels vary in size.

The two primary variants of ISDN are BRI (Basic Rate Interface) and PRI (Primary Rate Interface). BRI, sometimes referred to as 2B+D, provides two 64 Kbps B channels and a 16 Kbps D channel over a single 192 Kbps circuit (the remaining bandwidth is used for framing). BRI is the ISDN equivalent of a single phone line, though it can handle two calls simultaneously over its two B channels. PRI, essentially ISDN over T1, is referred to as 23B+D and provides 23 B channels and a 64 Kbps D channel. PRI is intended for use by an Internet Service Provider, for example, multiplexing almost two dozen calls over a single pair of wires. A number of international standards define ISDN. I.430 describes the Physical layer and part of the Data Link layer for BRI. Q.921 documents the Data Link protocol used over the D channel. Q.931, one of the most important ISO standards, documents the Network layer user-to-network interface, providing call setup and breakdown, channel allocation, and a variety of optional services. Variants of Q.931 are used in both ATM and Voice over IP (VoIP).

The two most important ISDN connection protocols are Q.931/32 and Q921. Q.921, also referred to as LAPD (Link Access Protocol—D Channel) and a close cousin of HDLC, is the Data Link protocol used over ISDN's D channel. We will discuss more on Q.931/32, because it is more involved with the application and exchange method on VoIP. Q931 is ISDN's connection control protocol, roughly comparable to TCP in the Internet protocol stack. Q.931 doesn't provide flow control or perform retransmission, because the underlying layers are assumed to be reliable and the circuit-oriented nature of ISDN allocates bandwidth in fixed increments of 64 Kbps. Q.931 does manage connection setup and breakdown. Like TCP, Q.931 documents both the protocol itself and a protocol state machine. The general format of a Q.931 message includes a single byte *protocol discriminator* (8 for Q.931 messages), a *call reference value* to distinguish between different calls being managed over the same D channel, a *message type*, and various *information elements* (IEs) as required by the message type in question. Figure 9.4 shows a typical Q.931 format.

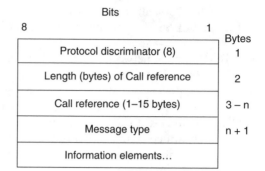

Figure 9.4

Q.931 format.

Q.931 has been designed for control signaling and thus is only used on the D channel, to establish, maintain, and release connections on B channels. It is a protocol between the user and the network. But where is the position of this protocol in the ISO/OSI reference model? As it only gives access to the network, it covers the network layer. Inside the switch there is an application involved that performs the translation between the D channel protocol and the SS7 protocol used for the signaling between switches, but for the subscribers putting up a circuit switch connection to the D channel protocol, it is transparent. Introducing Q.932 and supplementary services, the situation seems to become more complicated. In fact, there is still no layer 7 activity involved in Q.931. Introducing Q.932 doesn't change anything on the status of Q.931 in the reference model. Without Q.932 there is no possibility for interaction between an application and an exchange. It is, for example, not possible to disable the call waiting feature using only the means given by Q.931. This action would be not call-associated and only communication between switch and user equipment would occur—neither an end-to-end user connection nor any B channel is involved. But all messages defined in Q.931, require a valid call reference or an end-to-end connection. So a solution for this problem had to be found. And this solution is Q.932.

Q.932 becomes an integral part of Q.931 in the way that it adds new messages to the messages already known in Q.931, and furthermore it defines new information elements (FIE). From this point of view Q.932 is only an extension of Q.931. But on the other hand, Q.932 brings the part of layer 7 communication between applications—not known in Q.931—into the game.

So Q.931 has two parts. The first part is formed by the "classical" call control messages such as SETUP, DISCONNECT, etc. The second part is the extension formed by Q.932 that allows handling of layer 7 protocols by using the means of Q.931 plus the extensions provided by Q.932 to do so outside of a call context and a valid call reference.

SS7 over IP

Signaling System Number 7 (SS7) is a sophisticated, integrated signaling protocol designed to increase the efficiency of telephone service delivery. It has become the central nervous system for the delivery of wireline and wireless, and more recently IP services, the latest step in the convergence of communication networks. All networks need signaling to create connections, activate service, and deliver traffic. SS7 signaling can now be transmitted over an IP network by using the IETF SIGTRAN protocols. This enables operators to rationalize network operations and maintenance, in that dedicated signaling networks are no longer needed. SS7 over IP can also be used as a complement to existing signaling network, for example, for long-haul communications and when extremely high capacity is required.

SIGTRAN is a new IETF standard that initially focused on communications between media gateway controllers and signaling gateways. It has since developed into a more generic approach for transmitting signaling over IP networks. SCTP (Stream Control Transport Protocol) within SIGTRAN ensures reliable signaling transmissions over IP and solves deficiencies in TCP. M3UA (MTP3 User Adaptation Layer) is over SCTP and supports the SS7 call signaling protocols ISUP and SCCP, permitting them to communicate within IP networks.

Signaling System 7 was introduced by AT&T in 1975 and approved by the world-wide standard bodies in 1980. SS7 basic functions are divided into four sections called Layers.

MTP (Message Transfer Part)—Provides a reliable transfer and delivery of signaling information across signaling networks.

SCCP (Signaling Connection Control Point)—Provides additional routing capabilities via SubSystem Numbers (SSNs). It also offers the capability of routing based on dialed digits or global title translation.

ISUP (ISDN User Part)—Provides the transport of call set-up information between two signaling points.

TCAP (Transaction Capability Application Part)—Provides the capability of transferring non-circuit-related information between signaling points.

These basic protocols determine the ability to deploy applications worldwide and are used by additional higher-level protocols such as IS-41, ANSI-41, GSM MAP, AIN, and INAP. Compliance to standards determines the general applicability of the protocol. Strict compliance to the very latest standard revision is seldom possible. Compliance tables are used to precisely define a protocol capability for a given release. Do not overlook the importance of the programming language and operating system supported. These determine the tools and skills needed to develop your applications. An API that is the same architecture for all protocols and is operating system independent allows your applications to grow and move with little or no change. By making maximum use of valuable training, tools, and experience, the breadth of the operating system and protocol support determine the ease with which future applications and enhancements can be quickly accommodated. Figure 9.5 depicts the diversity of solutions achieved by using signaling transport protocols. Using SIGTRAN protocols such as an MTP3 user application (M3UA) and a signaling connection control part user application (SUA), the application vendor (i.e. Short Message service center [SMSC], IP-home location register [IP-HLR], and so on) only has to develop the application layer and does not have to deal with the complex SS7 interfaces.

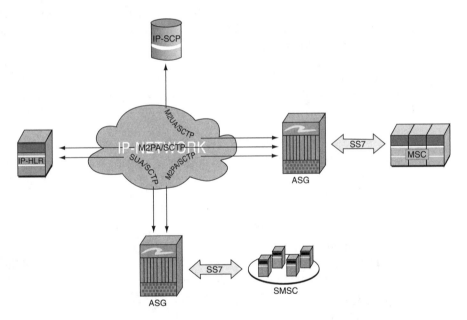

Figure 9.5

Sample implementation for signaling transport over IP.

SS7 supports API and different types of applications. When the application is mission critical, the software that is part of the API and protocol support infrastructure becomes very important. Reliability of the API and protocol support infrastructure must be part of the application design. This is often a larger effort than the application itself. Using fault-tolerant hardware does not mitigate the benefits of a complete, robust software infrastructure designed specifically for SS7 requirements.

With SS7, each protocol layer has areas of specification that are determined by the particular implementation of the protocol. Implementation limitations can have serious consequences. Details are very important, and the "broadest" implementations offer the best foundation for future applications and deployments. ISUP, unlike TCAP, has more than 100 country-specific accommodations. These variants ensured compatibility with existing circuit switches as SS7 network signaling was implemented.

Consideration of the SS7 network physical connection and associated hardware (board) capabilities is as important as selecting the protocols that support your application. The board determines the chassis you use, the number of SS7 links that can be connected, the manner in which maintenance is accomplished (online, offline), and whether the board is tested and delivered with the protocol software. A board that is fully integrated with the SS7 software is most desirable.

SS7 was originally established to define the procedures for the set-up, on-going management, and clearing of a call between two users of the Public Switched Telephony Network (PSTN). The scope of SS7 extends to both the network architecture and the protocols, which run on this network. More recently, SS7 has been used to provide signaling within IP networks to set up VoIP calls and even to manage calls that use a mix of PSTN and IP resources. By combining the SS7 and IP networks, services from one domain become available in the other (for example, Internet Call Waiting).

In a typical scenario, both telephone subscribers are connected to a local telephone exchange, referred to as a Central Office (CO). The SS7 SSP is part of this CO. When a subscriber makes a call, the SSP uses a protocol called ISUP to send messages to the called party's SSP in order to set up a voice circuit (or "trunk") between the two. Once ISUP has established the trunk, the two parties may begin their conversation. In some calls, the telephone number dialed by the caller may require translation before ISUP is able to identify the called party (for example, an 800 number must be translated into a physical telephone number). In some other cases, additional information may be passed with the called number (such as calling card number) that may have an effect on the processing of the call. In these cases, a protocol called TCAP is used by the calling party's SSP to access a network database that stores this information. Such a database is reached via an SCP node.

In the case where a call involves an IP network (for example, the called party is using an IP phone), the SS7 messages will pass through a Signaling Gateway to a Media Controller. The MCG uses the SS7 information to complete the call in the IP network. Operators want to migrate toward IP-based network architecture. Because the transition to an all-IP network will not happen overnight, both traditional circuit-switched and IP-based services need to be supported by a single network infrastructure simultaneously. It is believed that circuit switching will live for many years together with IP services. Hybrid architecture may be the best solution for most of the operators because it allows low-risk evolution from the current networks, while enabling a new services offering. These reasons have lead to several standard working-group studies, among them the SIGTRAN working group in the IETF organization. The architectural model that has been defined by the SIGTRAN working group enables the network evolution toward an all-IP network. The architectural model defines two new components of SCTP and several user adaptation sub-layer protocols (e.g. M2PA, M3UA, SUA), which together enable the required mechanisms for converging these two networks.

Transport—RTP, RTCP

RTP, the Real Time Transport Protocol, is used to provide end-to-end network transport functions suitable for applications transmitting real-time multimedia data such as audio, video, or data. The mode of transmission can be unicast or multicast type. The data transport is monitored by a sister protocol called Real Time Control Protocol (RTCP), which allows monitoring of the data delivery in a manner scalable to large multicast networks and provides minimal control and identification functionality. RTP was proposed by IETF in the RFC 1889. Incidentally, RTP is accepted as a universal standard for real-time multimedia transmission. Both ITU-T and the IETF have standardized this protocol in their systems. RTP is needed for real-time applications. The Transport Control Protocol (TCP) cannot support real-time services such as interactive video, conferencing, etc. The reason behind this is the fact that TCP is rather a slow protocol, requiring three-way hand shaking. Hence, UDP is used over IP as a better option over TCP and over IP. But UDP inherently is an unreliable protocol, which does not support retransmissions upon packet loss. Still, UDP has some features, such as multiplexing and check sum services, that favor real-time services. To encounter the drawbacks of UDP, RTP is proposed at the application layer level. The various services offered by RTP include payload type identification, sequence numbering, time stamping, and delivery monitoring. RTP can sequence those packets that arrive out of order at the receiving end. The sequence number can also be used to identify the lost packets. Time stamping is used for the proper playout of the media. The data arriving is continuously monitored by RTCP, which then informs the RTP layer to adjust its coding and transmission parameters for the proper delivery of the data. For example, if the RTCP layer detects severe packet loss, it may inform the RTP layer to slow down the rate of transmission. Though RTP aids in the proper playout of real-time media, please note that RTP does not, by itself, provide any mechanism to ensure timely delivery or provide other quality-of-service guarantees. It relies on lower-layer services to do so. RTP also does not guarantee delivery or prevent out-of-order delivery, nor does it assume that the underlying network is reliable and delivers packets in sequence. RTP is primarily designed to satisfy the needs of multiparticipant multimedia conferences. It is also applicable for services such as storage of continuous data, interactive distributed simulation, active badge, and control and measurement applications. The RTP protocol "provides end-to-end delivery services for data with real-time characteristics." This does not restrict the use of the protocol to multimedia data transport but expands its use to other fields, such as control and measurement, for example. This control connection is held over a separate channel (uses another port number than the RTP one).

The following are examples of applications for RTP. They depict the way the protocol works.

For audio conference, one of the following setups is required:

- A network with full multicast capability, or
- A server that accepts connections from all the participants in the conference and distributes the packets originated by one of those participants to all the rest.

Regardless of the setup used, the connection is established and the two ports are used by each member of the group: one for RTP packets and one for RTCP packets (similar to the way FTP works, with a data connection and a control connection). Once this is done, the audio coding scheme to be used is negotiated between the members, and the conference can begin. Periodically, each one of the members broadcasts a control packet to give information to the other participants about the connection and the quality of reception. If communication problems are found, the group can renegotiate the communication parameters. This can also be done if a new member with access through a constrained bandwidth channel enters the conference. When any of the participants wishes to leave the conference, he transmits a special goodbye message in the control channel.

An audio/video conference works much like the voice-only conference with the following differences:

- Instead of two data ports for each participant, three must be provided: one for control (RTCP messages), one for audio data, and another for video data.
- The data format and stream speed for the video data must also be negotiated.
- Because the video and audio data streams are separated, there must a means of synchronizing the two streams so they can be simultaneously played. This is done by using the timestamps each RTP packet carries. Reading the

timestamps for both data streams and using buffering techniques, the software can play back both data streams simultaneously.

The concept of network IP telephony is very similar to the one of audio conference, but a telephone conference is normally held between two people.

Telephony service normally requires a server that stores the network addresses of all the potential users. This server often has a directory service, used to easily find a particular user. Once a user decides to talk to another one, he uses the directory service to find the address of the remote user, and instructs the server to place a call.

The server tries to establish contact with the remote user; if it succeeds, then it helps the two parties negotiate the audio and/or video formats to use and lets the conference begin. While the teleconference is taking place, the server monitors both parties for quality of service, and also gives "busy signals" to other parties that try to establish communication with the ones that are already talking.

Once one of the parties issues a disconnect signal, the server frees the resources and the connections and advertises the users as "not busy."

Using special hardware, the server can also act as a "net-to-POTS" relay, enabling network users to talk over the network with users of the ordinary telephony service. The services provided by RTP are the following:

1. Payload type identification: Lets applications using the protocol know which kind of information is being transmitted (audio, video, etc).
2. Sequencing: When RTP runs on top of a datagram-based service, packet sequence numbers are required to reassemble the information stream in the destination (packets may not arrive in the correct order because each one can take an independent route from source to destination).
3. Timestamping: In applications such as teleconference, timestamps must be provided in the packets to assure synchronization of the devices participating in the conference; otherwise, information might be lost. This timestamping is also important in controlling and monitoring applications to have an accurate recording of the monitored events.
4. Delivery monitoring: Allows the participating applications to collect and share statistics about the performance of the data transport service.
5. Information about connection: Besides the statistics, more useful information may be distributed among the participants in a communication group, e.g., the names and locations of the parties, when and why a party is leaving, etc.

The last two services are provided by RTCP, the companion control protocol for RTP. The following services are NOT provided by RTP:

- Quality of service: RTP relies on the lower-layer mechanisms of the communications stack for in-time delivery of the packets, and for reserving adequate bandwidth for the communication to take place.
- Reliability in packet delivery: RTP does not check for missing packets or out-of-order delivery. This must be done by the upper-layer protocols.
- Security: If a secure connection is needed, the encryption mechanisms must be implemented in another layer of the communications stack. However, the RTP packets have features that make their encryption easy.

RTP has the ability to use the multicasting mechanism in the networks that provide it, in order to establish one-to-many connections. In the Internet world, RTP runs normally on top of UDP and makes use of its multiplexing and checksum services. However, any lower-level protocol that is adequate for the application may be used. Another interesting characteristic of this protocol is that its definition is very open, so it's fully customizable to suit the needs of the application that will use it.

The header of the RTP packet has the following format:

V	P	X	CC	M	PT	Sequence number
Timestamp						
Synchronization source (SSRC) ID						
Contributing source (CSRC) IDs . . .						

The header fields have the following meaning:

Version (V): The field is of length 2 bits, indicating the current version of RTP. The current version of RTP is 2.0.

Padding (P): This field is of length 1 bit. If P is set, the packet contains one or more additional padding octets at the end, which are not part of the payload. Padding is needed by some encryption algorithms, which require fixed block sizes, or for carrying several RTP packets in a lower-layer PDU.

Extension (X): This field is of length 1 bit. If X is set, the fixed header is followed by exactly one header extension.

CSRC count (CC): This field is of length 4 bits. The field indicates the number of CSRC identifiers that follow the fixed headers. As mentioned before, the field has a non-zero value only if passed through a mixer.

Marker bit (M): This field is of length 1 bit. If M is set, it indicates some significant events such as frame boundaries to be marked in the packet stream. For example, an RTP marker bit is set if the packet contains a few bits of the previous frame along with the current frame.

Payload type (PT): This field is of length 7 bits. PT indicates the payload type carried by the RTP packet. RTP Audio Video Profile (AVP) contains a default static mapping of payload type codes to payload formats. Additional payload types can be registered with IANA.

Sequence number: This field is of length 16 bits. The number increments by one for each RTP data packet sent, with the initial value set to a random value. The receiver can use the sequence number not only to detect packet loss but also to restore the packet sequence.

Time stamp: This field is of length 32 bits. The time stamp reflects the sampling instant of the first octet in the RTP data packet. The sampling instant must be derived from a clock that increments monotonically and linearly in time to allow synchronization and jitter calculations at the receiver. The initial value should be random, so as to prevent known plain text attacks. For example, if the RTP source is using a codec, which is buffering 20 ms of audio data, the RTP time stamp must be incremented by 160 for every packet irrespective of the fact that the packet is transmitted or dropped as silence.

SSRC: This field is of length 32 bits. This field identifies the source that is generating the RTP packets for this session. The identifier is chosen randomly so that no two sources within the same RTP session have the same value.

CSRC list: The list identifies the contributing sources for the payload contained in this packet. The maximum number of identifiers is limited to 15, as is apparent from the CC field (all zeros are prohibited in CC field). If there are more than 15 contributing sources, only the first 15 sources are identified. One observation that can be drawn from the RTP packet is that it does not contain the delimiting field, as is done in the lower-layer protocol data units (PDU). The reason behind this is that the payload of RTP is the same as that of the IP payload and hence not required. If the same user is using multiple media during a session, say for example, audio and video, separate RTP sessions are opened for each one of them. Hence there is no multiplexing of media at the RTP level. It is up to the lower layers to multiplex the packets from various media and send on a single channel. But RTCP maintains one identifier called CNAME, which is the same for all the media initiated by one user. Hence CNAME is the only identifier at the RTP layer level that can identify the media originated from a user. The overhead of RTP header is considerably large, as seen from the preceding. To reduce this, RTP header compression is proposed.

RTCP

RTP receivers provide reception quality feedback using RTCP report packets. The report packets can be of sender type or receiver type. The receiver sends SR (sender reports) if it is actively participating in the session. Otherwise it will send the receiver reports. In addition to the SR and RR packets, RTCP has other packet types: SDES (Source Description), BYE, and APP (Application defined). Table 9.1 shows the five types of RTCP packets and the value of the PT field for each.

Table 9.1

Packet Type	Abbrev	PT value
Sender report	SR	200
Receiver report	RR	201
Source description	SDES	202
Goodbye	BYE	203
Application-defined	APP	204

The RTCP sender report packet is as shown in Figure 9.6. The fields have the following meaning.

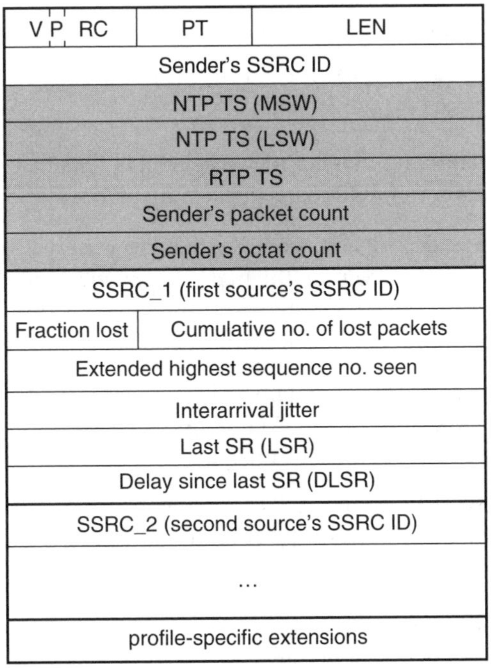

Figure 9.6

RTCP sender report packet.

Version (V): Same as in RTP.

Padding (P): Same as in RTP.

Reception report count (RC): This field if of length 5 bits. RC indicates the number of reception report blocks it contains.

Packet type (PT): This field is of length 8 bits. SR report packet has a packet type of 200.

Length: This field is of length 16 bits. The length of the packet is given in 32-bit words minus one.

SSRC: Same as in RTP.

NTP Time stamp: This field is of length 64 bits. NTP indicates the wall clock time when this packet is sent. This information is used to measure round-trip propagation delay.

RTP Time stamp: Same as in RTP. These time stamps are used for intra- and inter-media synchronization.

Senderís packet count: This field is of length 32 bits. This field indicates the number of RTP data packets transmitted by the sender from the start of the session up until this SR packet has been transmitted.

Senderís octet count: This field is of length 32 bits. It is the total number of payload octets sent from the beginning of the session.

Fraction lost: The fraction of RTP packets lost since the previous SR or RR packet is sent.

Cumulative number of packets lost: This field is of length 24 bits. It is the total number of RTP data packets lost from the beginning of the session.

Extended highest sequence number received: This field is of length 32 bits. The low 16 bits contain the highest sequence number received in an RTP data packet and the most significant 16 bits extend that sequence number with the corresponding number of sequence number cycles.

Inter-arrival jitter: This field is of length 32 bits. It is an estimate of the statistical variance of the RTP data packet arrival time measured in time stamp units and expressed as an unsigned integer.

Last SR timestamp: This field is of length 32 bits. It contains the middle 32 bits of 64 bits in the NTP time stamp received as part of the most recent RTCP SR packet.

Delay since last SR (DLSR): This field is of length 32 bits. It is the delay expressed in units of 1 in 65536 seconds, between the last SR packet and this RR packet.

RTCP SDES packet

RTCP SDES packet had different types of source description. The different types of source description packets are listed as follows:

CNAME: This is a canonical end-point identifier. CNAME is a mandatory item in the SDES packet, all the remaining being optional. CNAME can be used by the receivers to identify the different media streams originated from a particular user.

The other SDES items supported are NAME, EMAIL, PHONE, LOC, TOOL, NOTE, and PRIV.

There is a possibility that RTCP traffic can exceed the RTP traffic during a conference session involving larger numbers of participants. This may happen, because at any point of time, only one person will talk while the others listen. To reduce the RTCP traffic, the RTCP packet transmission rate is changed dynamically as a function of number of participants, RTCP packet size, RTCP bandwidth, etc. According to the standard, 20 percent of the session bandwidth is allocated to RTCP and 5 percent of the RTCP bandwidth is allocated to CNAME. In other words, RTCP RR and SDES packets are sent for every 5 RTP packets transmitted. For small sessions (conference with less number of participants), there would be an RTCP transmission for every 5 seconds. For every 15 seconds, one extra item would be included in the SDES packet.

RTCP BYE packet

The RTCP BYE packet is sent at the end of RTP session.

RTCP APP packet

The APP packet is intended for experimental use as new applications and new features are developed.

RTP and its companion protocol, RTCP, constitute a suite of open protocols for exchanging multimedia information in a network or an Internet. The protocol provides basic services for sequencing and reassembly of the packets, and basic quality of service metrics. Its applications include audio conferencing, video conferencing, VoIP, interfacing with ordinary telephone service, and more. RTP is highly dependent on lower-layer protocols for services such as quality of service assurance, bandwidth assignment, timely delivery of the packets, and security (privacy, authentication, and integrity). The use of mixers and translators allows RTP applications to overcome bandwidth limitations, protocol differences between participants, and security issues, among others. Since its creation, the RTP protocol has been improved by means of the definition of new payload types and new ways to deal with effects of data loss. Bandwidth conservation will always be an issue in transmission of multimedia data. A careful choice of coding algorithms for the data and the use of the RTCP bandwidth conservation mechanisms can make the system scale well, even for a large number of users. The openness of the protocols allows for improvements and additions to it, as new technologies and types of information arise.

CHAPTER 10

MANAGEMENT AND CONTROL PROTOCOLS

For converged network or managing real-time applications, it is important to understand real-time, call-level VoIP performance management and data network to improve service quality, maximize your existing network, and minimize operational costs. On a VoIP network topology, real-time voice monitoring (node to node) would be a very important feature on a converged management tool. To help understand the role of a network management protocol we need to first look at a network management model. A typical network management system is called a manager/agent model and consists of a manager, a managed system a database of information, and the network protocol. The manager provides the interface between the human network manager and the devices being managed. It also provides the network management process. The management process does such tasks as measuring traffic on a remote LAN segment or recording the transmission speed and physical address of a router's LAN interface. The manager also includes some type of output, usually graphical, to display management data, historical statistics, and so on. A common example of a graphical display would be a map of the internetwork topology showing the locations of the LAN segments. By selecting a particular segment the operator can display its current operational status.

The managed system consists of the agent process and the managed objects. The agent process performs network management operations such as setting configuration parameters and current operational statistics for a device on a given LAN. The managed objects include workstations, servers, wiring hubs, and communication circuits. Associated with the managed objects are attributes, which may be statically defined (such as the speed of the interface), dynamic (such as entries in a routing table), or require ongoing measurement (such as the number of packets transmitted in a given time period).

A database of network management information, called the Management Information Base (MIB), is associated with both the manger and the managed system. Just as a numerical database has the structure for storing and retrieving data, a MIB has a defined organization. This logical organization is called the Structure of Management Information (SMI). The SMI is organized in a tree structure, beginning at the root, with branches that organize the managed objects by logical categories. The MIB represents the managed objects as leaves on the branches.

Most devices on the network do not have the same information in their MIBs for two reasons. First, most devices usually come from different manufacturers who have implemented network management functions in different but complementary ways. Second, devices perform different internetworking functions, and may not need to store the same information. For example, a workstation may not require routing tables, and would not need to store routing-table-related parameters in its MIB. On the other hand, a router's MIB would not contain statistics such as CPU utilization that may be significant to a workstation.

The network management protocol provides a way for the manager, the managed objects, and their agents to communicate. To structure the communications process, the protocol defines specific messages, referred to as commands, responses, and notifications. The manager uses these messages to request specific management information, and the agent uses them to respond. The building blocks of the messages are called Protocol Data Units (PDU). For example, a manager sends a request PDU to retrieve information, and the agent responds with a response PDU.

SNMP for Network Management

SNMP is based on the manager/agent model. Its primary purpose is to allow the manager and the agents to communicate. This protocol provides the structure for commands from the manager, notifies the manager of significant events from the agent, and responds to either the manager or agent. The original version of SNMP was derived from the Simple Gateway Monitoring Protocol (SGMP) and was published in 1988. At that time the industry agreed that SNMP would be an interim solution until OSI-based network management using CMIS/CMIP became more mature. Since then, however, SNMP has become more popular than the OSI solution and has been more widely adopted than originally anticipated.

Overall, SNMP is designed to be simple. It does this three ways. First, by reducing the development cost of the agent software, SNMP has decreased the burden on vendors who wish to support the protocol. This increases the acceptance of the SNMP. Second, SNMP is extensible because it allows vendors to add their own network management functions. Third, it separates the management architecture from the architecture of network devices such as workstations and routers. This further widens the multivendor acceptance and support for this protocol.

SNMP is also referred to as simple because the agent requires minimal software. Most of the processing power and data storage resides on the management system, while a complementary subset of those functions resides in the managed system. To achieve its goal of being simple, SNMP includes a limited set of management commands and responses (see the SNMP Architecture diagram, Figure 10.1).

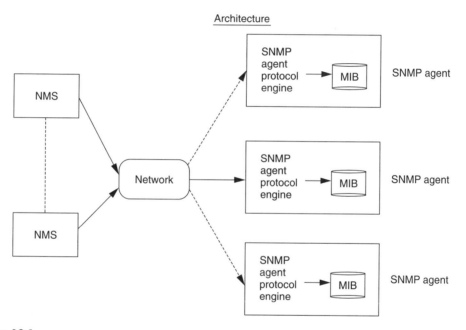

Figure 10.1

A typical SNMP architecture.

The management system issues Get, GetNext, and Set messages to retrieve single or multiple object variables or to establish the value of a single variable. Then the managed system sends a response message to complete the Get, GetNext, or Set. Finally the managed system sends an event notification, called a trap, to the management system to identify the occurrence of conditions such as a threshold that exceeds a predetermined value.

Because most management information does not demand the reliable delivery that connection-oriented systems provide, the communication channel between the SNMP manager and agent is connectionless. In other words, no pre-arranged communication path is established before the transmission of data.

Neither the manager nor the agent relies on the other for its operation. Consequently, a manager may continue to function even if a remote agent fails. When the agent resumes functioning, it can send a trap to the manager, notifying it of its change in operational status. Even though SNMP makes no guarantees about the reliable delivery of the data, in reality most messages get through, and those that do not can be retransmitted.

The most common implementation of SNMP is in the TCP/IP world. It relies on the User Datagram Protocol (UDP) and the Internet Protocol (IP) for proper operation. SNMP also requires Data Link layer protocols, such as Ethernet or Token Ring, to set up the communication channel from the management to the managed system.

An SNMP context is a collection of management information accessible by an SNMP entity. An item of management information may exist in more than one context. An SNMP entity potentially has access to many contexts. Typically, there are many instances of each managed object type within a management domain. For simplicity, the method for identifying instances specified by the MIB module does not allow each instance to be distinguished amongst the set of all instances within a management domain; rather, it allows each instance to be identified only within some scope or "context," where there are multiple such contexts within the management domain.

It is based on the OSI Network Management Architecture, but several aspects were simplified. SNMP explicitly minimizes the number and complexity of management functions realized by the management agent itself. This goal was selected for four main reasons.

1. The development cost for management agent software necessary to support the protocol is accordingly reduced.
2. The degree of management function that is remotely supported is accordingly increased, thereby admitting fullest use of Internet resources.
3. The degree of management function that is remotely supported is accordingly increased, thereby imposing the fewest possible restrictions on the form and sophistication of management tools.
4. Simplified sets of management functions are easily understood and used by developers of network management tools.

A goal of the architecture was that the base paradigm of monitoring and control could be extended to additional, possibly unanticipated, aspects of network operation and management. Also, the architecture goal was that the architecture be as independent as possible of the architecture and mechanisms of particular systems.

SNMP was from the start designed to support multiple administrative entities and inter-domain management. The administrative framework is based on management communities that share common views of the managed objects. This framework supports a trivial authentication mechanism, which in most cases can be easily broken, and as a consequence SNMP control should only be able to do limited damage to the operation of the network elements.

There are very few types of PDUs in the SNMP protocol.

1. GetRequest-PDU, to request the value of a management object
2. GetNextRequest-PDU, to allow the transversal of a list of management objects
3. GetResponse-PDU, to reply to a get-request or get-next-request
4. SetRequest-PDU, to change the value of a management object
5. SetResponse-PDU, to reply to a set-request
6. Trap-PDU, to signal an exceptional condition

Except for the Trap-PDU, all protocol interactions are initiated by the manager entity. Object information is encoded according to the Structure of Management Information (SMI) described in RFC1155, based on a subset of ASN.1. SNMP is a connectionless lightweight protocol on top of the connectionless transport protocol of the Internet Protocol Suite UDP. Security is trivial, based on a community string sent within each PDU header. Filtering, scoping, dynamic manipulation of objects, and action requests are put outside the scope of the protocol. MIBs for objects that need these constraints can include entries that implement these functionalities.

SNMP has a very straightforward architecture. The manager/agent applications reside at the Application Layer. Below the Application Layer is the Presentation Layer. The ASN.1 encoding resides at this layer and ensures uniform structure of management information and proper syntax for the SNMP messages. SNMP's connectionless communication mechanism removes some of the need for a Session layer and reduces the responsibilities of the lower layers. For most implementations, the UDP performs the Transport Layer functions and the IP provides the Network Layer functions, and LANs such as Ethernet and Token Ring provide the Data Link and Physical Layer functions.

The use of SNMP agents within internetworking devices has increased dramatically in the last few years. There are five general categories of devices in which you will find agents: wiring hubs, network servers and their associated operating systems, network interface cards and the associated hosts, internetworking devices such as bridges and hubs, and test equipment such as network monitors and analyzers. Other devices such as uninterruptible power supplies have also become SNMP compatible.

Using LDAP, COPS, and PBNM for Policy-Based Controls

Each day, companies strive to correlate business decisions to things that actually happen on their network, selecting which users have access to which network resources, prioritizing which applications are critical to company operations, delivering tiered bandwidth and differentiated services to each customer according to their needs, managing the voice, video, and data demands on corporate LAN and WAN links, and managing the overall flow of traffic through internal and external networks. All of these actions share a common requirement by applying corporate business policies to specific network actions including bandwidth management, firewalling, caching, routing, and VPN equipment. Today, the network manager or network service provider must manage policies for all of these devices by managing a wide array of users, applications, and resources to determine policies and then configuring each individual piece of equipment.

In the traditional network, new services are developed, updated, and deployed slowly by adding new hardware. The physical network is then reconfigured to reflect the additions. Network administrators (net admins) in these environments are accustomed to managing their networks by manually configuring one piece of equipment, then manually configuring the next piece of equipment, and so on. In this inflexible paradigm, traditional standards for switches, routers, and other network devices are designed for manual interaction, not for fast and automated machine-managed interaction. The traditional network system is not only slow to change and inefficient, but it is also difficult to protect, scale, and upgrade, and that's a problem. What is needed is a comprehensive, policy-based system that will allow the network manager to define, in a succinct and organized fashion, corporate policies that automatically effect change on specific equipment in the network environment. The end result is that the end-to-end network performance will meet the general expectations of the corporate, commerce, or service provider environment.

Business decisions, more and more, affect the entire network infrastructure. To manage today's business, the network must be reliable. The network must be predictable. To accomplish this, an overall end-to-end strategy must be developed to correlate the business with the overall network actions. This policy-based network architecture document details some of the goals and processes involved in developing this system. Allot's policy-based network system delivers a comprehensive architecture that allows the merging of high-level user, application, and resource policy information with network-wide policy actions.

The solution to this problem has four elements.

Interface to Directory Database, LDAP

This provides access to higher level user and application-level policy information via data stored in common directory repositories. This high-level database information can then be translated into actual policy enforcement actions. The

Directory Database Server provides the central repository for policy information. In general, the Directory Database resides on a Directory Server as a third-party LDAP-based application provided via companies delivering directory-related products such as Avaya, Microsoft, Netscape, IBM, and Novell. The Directory Server is responsible for the storage of a wide range of information including e-commerce, user login, and network yellow pages as well as network-specific policy information. The directory can be grouped in a distributed and hierarchical fashion for sharing data. The key to the directory is the wealth of applications that allow users to organize and correlate the wealth of information that subsequently gets stored into the database. A user can, for example, enter their entire customer list into the directory and then assign specific administrative and policy information on each customer including network-specific bandwidth policies. The directory will interface to the Policy Decision Administrator in order to create higher-level, dynamic policies. An administrator will be able to define a customer as "gold level" and then set up a specific policy that says that all gold-level customers will receive a specific priority throughout the network. Advanced features will allow the defining of dynamic policies that will allow network managers to script specific application conditions in order for a network action to take place. A network administrator, for example, would be allowed to assign a given priority to a customer who spent a minimum amount of money in the last year.

The Data Abstraction Layer is the interface to the directory. It maintains and has knowledge of the policy structure and content. In its basic form, it allows abstract definition of policies defined in the directory to be translated to physical characteristics in the network. A manager, for example, can define a concept of "Service Type" in the directory. He can then set a given *customer* in the directory to have a *Service Type* of, for example, *Gold*. The Data Abstraction Layer then provides pre-defined and user-defined variables that define things such as

- *Service Type* is an LDAP directory schema variable.
- *Service Type* translates, in this case, *customers* into a concept an enforcement device can understand, namely a group of IP addresses.
- A new entry called *Service Type* will now be accessible through the Policy Manager User Interface. It will serve as a new *Rule Condition* (along with things such as IP addresses, protocol types, and time-of-day).
- The manager can now declare, using the Policy Manager User Interface, that, as a condition, a *Service Type* equal to *Gold* will translate to a specific class-of-service. Perhaps, in this case, *Gold* would specify a minimum bandwidth of 128Kbytes/sec and a maximum burstable bandwidth of 256Kbytes/sec.

The device abstraction layer translates policies to specific device schemas and then relays this information to specific policy enforcement devices. For the near-term, there will be different types of enforcement products that will communicate with the Device Abstraction Layer in different formats and protocols. This may include protocols and architectures such as COPS, LDAP, SNMP, or, for some legacy products, a command-line interface.

QoS Gateway

This is responsible for providing end-to-end policy enforcement and management via standards-based signal provisioning protocols including Differentiated Services, ToS, RSVP, MPLS, and 802.1P. The Policy Enforcer can be a simple router that makes policy decisions based on a field in a particular "tagged" packet. Alternatively, the Policy Enforcer may be a piece of equipment that locally consolidates and analyzes traffic flows and network conditions in order to perform complex network actions such as

- *Traffic Conditioning and Shaping.* This includes things such as traffic prioritization, traffic guarantees, and bandwidth management.
- *Policing.* This includes access control, user authentication, and remote login.
- *Tagging/Signal Provisioning.* This includes translating and relaying signal provisioning information (RSVP, Differentiated Services, 802.1P, MPLS) through the network.
- *Server Resource Control.* This includes advanced enforcement capabilities such as server load-balancing and cache redirection.

Typically, this more complex piece of equipment will sit at the edge or border of a network domain and relay, via signaling protocols, overall bandwidth requirements to the internal network equipment. When referring to Policy Enforcement, this document refers primarily to these edge devices and, specifically, to the Allot AC Policy Enforcer System itself.

The policy-based management for QoS functions is also responsible for coordinating administrator input with other external and internal policy information and translating this information into concise network terminology. The Policy Management service is responsible for coordinating the flow of bandwidth between various devices that make up the input and output to specific domains. By managing the critical border locations in a specific domain and then having the enforcement devices communicate, via standard signaling protocols, to the internal devices, this system creates a complete end-to-end policy-based network solution. Policy-based control functionality can be distributed and hierarchical in structure, representing various physical and logical domains for management of specific Policy Enforcement devices. As an example, an enterprise domain can be defined for the whole corporation, including general policies and management structure that affect every user, application, or resource. Individual departments "inherit" the higher-level corporate policies but can then define their own enforcement policies for the clients and servers that they control. The corporate network administrator can delegate administration rights to the local department's administrator.

Policy-Based Network Management (PBNM)

Centralized, automated, and coordinated management of network services via a whole-network perspective, PBNM (policy-based network management) enables the coordinated control of network services. PBNM solutions are different from general device-management solutions in that PBNM is about network-wide policy and customer requests for services, not per-device configuration. PBNM lets net admins automate most or all of the traditionally manual device management and configuration tasks. Basically, PBNM lets them create higher-level user and customer policies instead of setting detailed queue configurations based on machine addresses and obscure port numbers. With PBNM, net admins can configure devices in the context of the network as a whole. PBNM automates the control of the network infrastructure by storing policies on and distributing policies from servers.

The high-level policies themselves are described using an if-then approach: If certain conditions are present, then specific actions are taken. "If" conditions could include a time of day, type of traffic, IP address, person, group, etc. "Then" actions could include implementing priority tagging, setting security encryption to a certain level, or could relate to actions such as access control and load balancing. For example, a new service might be needed to deal with an emerging type of network threat, such as a new denial-of-service attack. This could require a change in the way blocks of traffic are scheduled, prioritized, and policed. The net admin makes the necessary changes to the existing policies or installs software that implements a new network-wide denial of service protection scheme. Those changes are then automatically interpreted and propagated throughout the network to all appropriate programmable devices, via the policy servers. The programmable devices needed to address the threat are dynamically reconfigured as directed by the newly defined policy rules.

Likewise, broker processes can automatically field user, application, and customer requests for network services. These services brokers interact with the appropriate PBNM components. They reconfigure the appropriate network devices to dynamically grant the requested service as afforded by the current administrative policies and network conditions. Such a system lets ISPs extract value from networks for new services. It also allows peer networks to automatically vie for services between one another, so that even limited networks can offer services end-to-end. With PBNM, the net admin doesn't have to manually reconfigure every switch or router on a data path to grant a new level of service. Instead, the changes can happen automatically and in real time, based simply on a set of standardized business rules and user or application requests. This is where the new, integrated standards come into play.

Common Open Policy Service

COPS standards tie PBNM and the programmable network processors into the new paradigm. The COPS protocol was developed in the Resource Allocation Protocol (RAP) Working Group of the IETF. The COPS protocol is a protocol for exchanging network policy information between a policy decision point (PDP) in a network and policy enforcement points (PEPs) as part of overall Quality of Service (QoS)—the allocation of network traffic resources according to desired priorities of service. The policy decision point might be a network server controlled directly by the network administrator who enters policy statements about which kinds of traffic (voice, bulk data, video, teleconferencing, and so forth) should get the highest priority. The policy enforcement points might be router or layer 3 switches that implement the policy choices as traffic moves through the network. Currently, COPS is designed for use with the Resource Reservation Protocol (RSVP), which lets you allocate traffic priorities in advance for temporary high-bandwidth requirements (for example, video broadcasts or multicasts). It is possible that COPS will be extended to be a general policy communications protocol.

In operation, RSVP makes two determinations when an RSVP request arrives at a router or layer 3 switch. First, it determines whether there are enough resources to satisfy the bandwidth reservation request. If there are, RSVP determines whether the user is authorized to make the reservation. The first determination is known as the admission control decision; the second is known as the policy control decision. COPS allows the router or layer 3 switch to communicate with the policy decision point about whether the request for the bandwidth reservation should be permitted. Without COPS, all resources would be reserved on a first-come first-served basis only, and one or more requesters could easily take all the bandwidth.

The protocol employs a client/server model where the PEP sends requests, updates, and deletes to the remote PDP and the PDP returns decisions back to the PEP; i.e., the PEP is the client, the PDP is the server.

The protocol uses TCP as its transport protocol for reliable exchange of messages between policy clients and a server. Therefore, no additional mechanisms are necessary for reliable communication between a server (PDP) and its clients (PEPs). The protocol is extensible in that it is designed to leverage off self-identifying objects and can support diverse client-specific information without requiring modifications to the COPS protocol itself. The protocol was created for the general administration, configuration, and enforcement of policies.

COPS provides message-level security for authentication, replay protection, and message integrity. COPS can also reuse existing protocols for security such as IPSEC to authenticate and secure the channel between the PEP and the PDP.

Request/Decision state is shared between client and server i.e., requests from the client PEP are remembered by the PDP until they are explicitly deleted by the PEP. States from various events (Request/Decision pairs) may be inter-associated. At the same time, decisions from the remote PDP can be generated asynchronously at any time.

Each COPS message consists of the COPS header followed by a number of typed objects (see Figure 10.2).

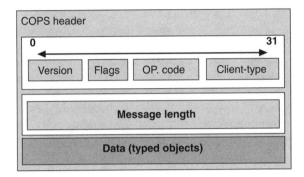

Version .4 bits.
COPS version number.

Flag .4 bits.
0x1 Solicited Message Flag Bit
This flag is set when the message is solicited by another COPS message.

Opcode. 8 bits.

Opcode	Description
1	REQ, Request.
2	DEC, Decision.
3	RPT, Report State.
4	DRQ, Delete Request State.
5	SSQ, Synchronize State Req
6	OPN, Client-Open.
7	CAT, Client-Accept.
8	CC, Client-Close.
9	KA, Keep-Alive.
10	SSC, Synchronize Complete.

Client type. 16 bits.
Identifies the policy client. Interpretation of all encapsulated objects is relative to the client type. Client types that set the most significant bit in the client-type field are enterprise specific. For KA Messages, the client-type in the header MUST always be set to 0 as the KA is used for connection verification (not per client session verification).

Figure 10.2

COPS message.

Client type	Description
1	RSVP

Message length. 32 bits.
Size of the message in bytes, which includes the standard COPS header and all encapsulated objects. Messages MUST be aligned on 4 byte intervals.

Data. Variable length.
Contains one or more self-identifying Objects.

Figure 10.2

Continued

Using the COPS protocol, developers can easily describe new capabilities for network devices in the context of the even larger, more complete services framework already provided by COPS. Under this framework, network devices can be automatically reconfigured by remote processes to implement new network services, enforce updated administrative policies, or handle customer requests for network services on the fly. At this time, the foundations of these new standards have already been laid. This is because key components of the foundation (including the COPS protocol and its support for QoS) have already become Internet Standards. Extensions to the COPS framework for controlling DiffServ are also already well developed, as are a number of other extensions, including feedback mechanisms, integrated access control, and traffic engineering. Likewise, developers can easily write their own COPS extensions (called PIBs, for Policy Information Bases) using the SPPI (Structure of Policy Provisioning Information) language. This lets COPS control and monitor any new network processor building-block capability, already conceived or yet to be conceived, as an extension to the existing framework. Along with the PBNM, this new standard and its extensions give developers a robust, extensible, dynamic framework to easily implement advanced network services.

MGCP and External Call Control Elements

Media Gateway Control Protocol (MGCP) is a protocol used for controlling Voice over IP (VoIP) gateways from external call control elements. MGCP is the emerging protocol that is receiving wide interest from both the voice and data industries. MGCP defines a protocol for controlling Telephony Gateways from external call control elements called Media Gateway Controllers or Call Agents. MGCP is central to VoIP solutions and may be integrated into products such as Central Office Switches, Gateways, Network Access Servers, Cable Modems, PBXs, etc., to develop a converged voice and data solution.

The evolution of the MGCP specification was largely influenced by "political" conflicts between proponents of alternative proposals for decomposed gateway architectures. In particular, MGCP owes its origin to the confluence of the

SGCP (Simple Gateway Control Protocol) and IPDC (Internet Protocol Device Control) protocols. MGCP is a complementary protocol to both SIP and H.323. It was designed specifically as an internal protocol between MGCs and MGs for decomposed gateway architectures. In the MGCP model, an MGC handles call processing by interfacing with the IP network via communications with an IP signaling device such as an H.323 gatekeeper or SIP Server and with the circuit-switch network via an optional signaling gateway. Using an H.323 analogy, the MGC implements the "signaling" layers of H.323 and presents itself as an "H.323 Gatekeeper" or as one or more "H.323 Endpoints." Within the MGCP approach, MGs focus on the audio signal translation function, performing conversion between the audio signals carried on telephone circuits and data packets carried over the Internet or other packet networks.

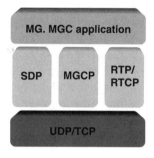

Figure 10.3

A sample architecture of MGCP I.

MGCP-compliant VoIP gateways provide a highly scalable solution for global IP communications networks and will co-exist in networks with other H.323- and SIP-compliant entities.

MGCP is a master/slave protocol with a tight coupling between the MG (endpoint) and MGC (server). Like SIP and H.323, it relies on a variety of other existing protocols such as SDP for describing the media aspects of a call, and RTP/RTCP (used by MGs) for handling the real-time transport of media streams.

In an MGCP architecture, the MGC server or "call agent" is mandatory and manages the calls and conferences and supports the services provided. The MG endpoint is unaware of the calls and conferences and does not maintain call states. MGs are expected to execute commands sent by the MGC call agents. MGCP assumes that call agents will synchronize with each other, sending coherent commands to MGs under their control. MGCP does not define a mechanism for synchronizing call agents.

MGCP main entities are Endpoints and Connections. MGCP assumes a connection model where the basic constructs of Endpoints and Connections are used for establishing voice paths between call participants. Endpoints are sources or sinks of data and can be physical or virtual. Physical endpoint creation requires hardware installation, while software is sufficient for creating a virtual endpoint. An interface on a gateway that terminates a trunk connected to a PSTN switch is an example of a physical endpoint. An audio source in an audio-content server is an example of a virtual endpoint.

Connections may be either point-to-point or multipoint. A point-to-point connection associates two endpoints. Once this association is established for both endpoints, data transfer between these endpoints can begin. A multipoint connection is established by connecting the endpoint to a multipoint session.

Connections can be established over several types of bearer networks: audio packet transmission using RTP and UDP over a TCP/IP network, audio packet transmission using AAL2 or another adaptation layer over an ATM network, and transmission of packets over an internal connection. For both point-to-point and multipoint connections the endpoints can be in separate gateways or in the same gateway.

The control primitives for MGCP operations are Signals sent from the MGC to the MG, and Events sent from the MG to the MGC. The concepts of Signals and Events are used for establishing and tearing down calls. Operations are performed by applying Signals TO, and detecting Events FROM endpoints. An MGC initiates transactions to manage/configure MG endpoints using MGCP commands. MGs send responses to MGC transaction requests using either a notification or restart command.

MEGACO/H.248, the New Standard

MEGACO is a general-purpose gateway-control protocol standardized in the IETF as RFC 3015 and as recommendation H.248 in the ITU-T. It is a master–slave, transaction-oriented protocol in which Media Gateway Controllers (MGC) control the operation of Media Gateways (MG). The MEGACO Stack and other Voice over Packet (VoP) software solutions from Hughes Software Systems (HSS) provide a complete set of building blocks for quickly developing next-generation solutions. The solutions meet the carrier class requirements of redundancy, scalability, and reliability.

MEGACO/H.248 is central to the VoIP solutions and may be integrated into products such as Central Office Switches, Gateways (trunking, residential, and access), Network Access Servers, Cable Modems, PBXs, IP Phones, Softphones, IADs, Middleboxes, etc. to develop a convergent voice and data solution.

MGCP was itself created from two other protocols, Internet Protocol Device Control (IPDC) and Simple Gateway Control Protocol (SGCP). The MGCP specifies a protocol at the Application layer level that uses a master–slave model, in which the media gateway controller is the master. MGCP makes it possible for the controller to determine the location of each communication endpoint and its media capabilities so that a level of service can be chosen that will be possible for all participants. The later Megaco/H.248 version of MGCP supports more ports per gateway, as well as multiple gateways, and time-division multiplexing (TDM) and asynchronous transfer mode (ATM) communication.

The Megaco/H.248 Protocol represents a single gateway control approach covering all gateway applications. This includes PSTN trunking gateways, ATM interface, analog line and phone interfaces, Internet telephones, announcement servers, and many more. It is this breadth of supported applications, along with key attributes of simplicity, efficiency, flexibility, and cost effectiveness, that makes Megaco/H.248 a compelling standard for use in next-generation networks.

The MGC layer contains all call control intelligence and implements call-level features such as forward, transfer, conference, and hold. This layer also implements any peer-level protocols for interaction with other MGCs or peer entities, manages all feature interactions, and manages any interactions with signaling such as SS7. The MG layer implements media connections to and from the packet-based network (IP or ATM), interacts with those media connections and streams through application of signals and events, and also controls gateway device features such as user interface. This layer has no knowledge of call-level features, and acts as a simple slave. The Megaco/H.248 Protocol drives master/slave control of the MG by the MGC. It provides connection control, device control, and device configuration. Because the Megaco/H.248 Protocol

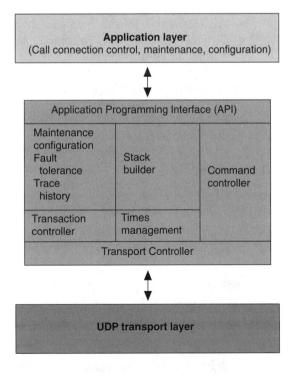

Figure 10.4

MEGACO/H.248 Stack.

is separate from and independent of the peer call control protocol (e.g., SIP or H.323), different systems can be used at the call control level with minimal cost impact on the gateway control layer.

MEGACO/H.248 Stack: Functions

- Association management: Creation and management of control association between MGC and a MG
- Transaction and Command management: Transaction and Command requests and replies management
- ALF support: Ensures reliable delivery of transactions
- Encoding/Decoding in both text and binary
- IN and ATM SDP encoding support
- Easy Packages upgrade support (validation, addition, and editing of packages)

MEGACO/H.248 Stack: Service User Interface

The Service User Interface exposes Protocol APIs and Management APIs to the service user. It provides the following functions to the service user:

- Interfaces to:
  - Register with the stack
  - Configure the protocol parameters
  - MG/MGC application for sending protocol messages to the MGC/ MG application
  - Activate and deactivate management features of the stack
  - Protocol trace, statistics collection, status reporting, error reporting, and redundant mode operation
- Flexibility to collate commands into transactions
- Protocol message encoding and decoding services
- Parameter validation services including package parameter validation. The interface may operate in the same process context (functional interface) or a different context (message-based interface) than the core stack

Redundancy Support

The MEGACO/H.248 stack provides multiple redundancy configurations to support different applications. Support options include

- Redundant applications
- Redundant MEGACO/H.248 stacks

Scalability

The MEGACO/H.248 stack design offers capability for scaling as the target application grows. Scalability can be achieved by the following means:

- Increasing the number of call agents and MG-MGC associations, which the stack handles as the system capacity grows

- Moving from uni-processor environment to multi-processor target hardware
- Moving from a single thread of control operating system to a multi-tasking real-time operating system
- Init time dimensioning of parameters

MEGACO/H.248 Operation and Management Support

- Extensive statistics collection
- Error reporting
- Status reporting
- Multi-level tracing support

Megaco/H.248 and MGCP comparison

Megaco/H.248 Protocol, the industry accepted open standard, is a fully developed and feature-rich approach that provides numerous options for building added-value products with differentiated features. MGCP, in comparison, is a closed approach that does not reflect true industry consensus and, therefore, has limited potential for multi-vendor interoperability. MGCP has serious technical flaws relative to Megaco/H.248.

	Megaco/H.248	**MGCP**
Standardization	• A truly open standard • Subject to review and compromise • Single international standard from both IETF and ITU	• Closed (effectively proprietary) • Not subject to open review and compromise
Multi-vendor interoperability	• Thoroughly reviewed and rigorous • Few inconsistencies and assumptions • High multi-vendor interoperability potential, lower interoperability cost and risk	• Incomplete and informally defined • Many inconsistencies and assumptions • Lowering interoperability potential and increasing interoperability risk/cost
Connection model	• Totally flexible termination-context model, supports all types of networks • Efficiently handles complex connection scenarios (conference, consultation, music on hold, etc), multimedia mixing, etc.	• Less flexible connection model • Inefficient for conferences and other complex connection scenarios • Different service implementations for IP and

Table 10.1

Megaco/H.248 and MGCP comparison

	Megaco/H.248	**MGCP**
Resource model	• Applicable to all packet network types, same service design for both IP and ATM • Allows for simple lawful intercept (CALEA) support • Physical resource is isolated from packet connection • Allows for efficient handling of complex services such as announcements, call hold, and bridging	ATM networks, increased complexity, more difficult to deliver and deploy • Lawful intercept (CALEA) support unclear • Association between physical resource and packet connection • Manipulations not performed directly • Decreased real-time performance
Package extensibility mechanism	• Easy definition of new application interfaces through cleanly defined, fully open package definition mechanism and IANA registration process • New packages can be defined based on existing packages	• Difficult to extend monolithic design • No clean, open package definition mechanism • No clear methodology for package extension based on existing packages
Profile mechanism	• Allows package extension without affecting base protocol standard • Short time to market for new functionality, increased opportunity for innovation • Allows definition and implicit MGC–MG agreement on application-specific interworking assumptions • Reduces complexity and increases efficiency in both MG and MGC • Improved interoperability for the specific applications	• Not available

Table 10.1

Continued

MEGACO/H.248 is an incompatible next-generation protocol to MGCP. The IETF (Internet Engineering Task Force) group that developed MGCP has joined forces with the ITU (International Telecommunications Union) to develop a new version that will be standardized by both agencies. Most companies that are currently developing MGCP products plan to move to MEGACO when it becomes standardized and adopted by the industry. In the context of this paper, MGCP and MEGACO fill the same role.

There are several advantages of using MGCP and IP-based communications systems over traditional telephony engineering models. Among these are scalability, expandability, development time, reliability, and vendor independence. With an IP-based communications agent talking to multiple IP-based media gateways over an IP network, the number of ports you can install into the system is virtually limited only by the number of IP addresses available in the network. It could also be limited by the number of call setup requests per second an agent can support and how big a database of phone numbers and IP addresses, etc. an agent has to maintain. However, these limitations can be overcome by using higher-performance platforms, and by distributing an agent's functionality over multiple systems.

In most cases you will want to use SIP and/or H.323 or even SS7 to build an agent. These are the standards for interoperability with the rest of the world. MGCP is NOT a replacement for SS7, SIP, or H.323. Instead, it is a step toward decomposing the gateway. SS7, SIP, and H.323 negotiate calls between terminals and telephony systems, MGCP is used within a telephony system to allow the system to be decomposed into physically separate components. There could be some applications where you could only use MGCP. These would be closed systems such as PBXs or Switches where interoperability with other IP-based equipment is not desired.

SIGTRAN Suite of Protocols

Signaling Transport Internet Draft (SIGTRAN) was developed by the IETF working group SIGTRAN and defines the control protocol between the Signaling Gateway, Media Gateway Controllers, and IP-based Signaling Points. The protocol transports message-based signaling protocols messages, usually SS7, transparently over IP networks. This protocol suite is made up of a new transport layer—the Stream Control Transmission Protocol (SCTP) and a set of User Adaptation (UA) layers that mimic the services of the lower layers of SS7 and ISDN. This Internet Draft defines a protocol for the transport of any Signaling Connection Control Part-User signaling (e.g., Transaction Capabilities Protocol, Radio Access Network Application Protocol, etc.) over IP using the Stream Control Transport Protocol. The protocol should be modular and symmetric, to allow it to work in diverse architectures, such as a Signaling

Gateway to IP Signaling Endpoint architecture as well as a peer-to-peer IP Signaling Endpoint architecture. Protocol elements are added to allow operation between peers in the Signaling System No.7 and IP domains.

SIGTRAN is used for relaying signaling messages from lower layers through an IP network to upper layers centralized in a single Softswitch. SIGTRAN is also used for transporting signaling through IP by replacement of the transport part of a signaling network and keeping the upper layers without any change (UMTS Core Network).

SIGTRAN protocol has flexible configurability to contain M3UA, M2UA (back haul, peer-to-peer, or both), IUA, and/or SUA, according to the customer's requirements. High scalability of system resources and performance requirements through configuration of system parameters and high code efficiency are achieved through separate compilations for SGs and MGC stacks.

Such robust design of stack architecture offers:

High performance through zero buffering of payload

Easy portability to different operating systems through a logically separated OS layer

Minimal inter-functional block communication (minimal IPC overheads)

Highly efficient timer and memory management

The solid architectural design of the SIGTRAN stack enables its unsurpassed performance and conformance to the evolving interoperability standards. The three major components within the SIGTRAN protocol stack include (see Figure 10.5).

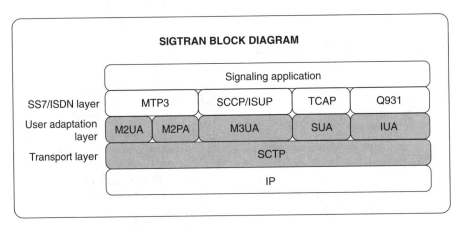

Figure 10.5

Sigtran Block Diagram.

- **User Adaptation Layer**—Receives primitives from the User Layer, adds adaptation-specific headers, identifies the destination, and routes the message over Stream Control Transmission Protocol (SCTP). It also receives messages from SCTP, removes adaptation-specific headers, and passes them on to the User Layer with a respective physical interface ID.
- **Common Signaling Transport Protocol Layer**—Supports a common set of reliable transport functions for the signaling transport.
- **Internet Protocol (IP) Component**—Serves as an unmodified IP transport protocol layer.

M2UA, M2PA, M3UA, SUA, IUA, and STCP are the new protocol layers as defined by SIGTRAN, while the grey areas illustrate existing protocols. ISUP is shown above, this would normally run over MTP3 or M3UA. The UA layers are named according to the service they replace, rather than the user of that service. For example, M3UA adapts SCTP to provide the services of MTP3, rather then providing a service *to* MTP3.

The SIGTRAN adaptation layers all serve a number of common purposes.

- To carry upper layer Signaling Protocols over a reliable IP-based transport.
- To provide the same class of service offered at the interface of the PSTN equivalent. For example, M3UA must provide the same look and feel to its users as MTP3, in terms of services, at least (M3UA does not actually replace the features and operations of MTP3).
- To be transparent: The user of the service should be unaware that the adaptation layer has replaced the original protocol (although this is largely dependent on the implementation).
- To remove as much need for the lower SS7 layers as possible.

SIGTRAN currently defines six adaptation layers, as follows:

1. *M2UA* provides the services of MTP2 in a client–server situation, such as SG to MGC. Its user would be MTP3.
2. *M2PA* provides the services of MTP2 in a peer-to-peer situation, such as SG-to-SG connections. Its user would be MTP3.
3. *M3UA* provides the services of MTP3 in both a client–server (SG to MGC) and peer-to-peer architecture. Its users would be SCCP and/or ISUP.
4. *SUA* provides the services of SCCP in a peer-to-peer architecture, such as SG to IP SCP. Its user would be TCAP, or another transaction-based application part.
5. *IUA* provides the services of the ISDN Data Link Layer (LAPD). Its user would be an ISDN Layer 3 (Q.931) entity.
6. *V5UA* provides the services of the V.5.2 protocol.

The framework is flexible enough to allow for the addition of new layers as required. The Stream Control Transmission Protocol (SCTP), which is a part of the SIGTRAN protocol suite, is designed to transport PSTN signaling messages over IP networks, but is capable of broader applications. SCTP is an application-level datagram transfer protocol operating on top of an unreliable datagram service such as UDP. It offers the following services to its users:

- Acknowledged error-free non-duplicated transfer of user data.
- Application-level segmentation to conform to discovered MTU size.
- Sequenced delivery of user datagrams within multiple streams, with an option for order-of-arrival delivery of individual datagrams.
- Optional multiplexing of user datagrams into SCTP datagrams, subject to MTU size restrictions.
- Enhanced reliability through support of multi-homing at either or both ends of the association.

The design of SCTP includes appropriate congestion avoidance behavior and resistance to flooding and masquerade attacks. The SCTP datagram is comprised of a common header and chunks. The chunks contain either control information or user data.

The following is the format of the SCTP header:

2 bytes	2 bytes
Source Port Number	Destination Port Number
Verification Tag	
Adler 32 Checksum	

Source Port Number

This is the SCTP sender's port number. It can be used by the receiver, in combination with the source IP Address, to identify the association to which this datagram belongs.

Destination Port Number

This is the SCTP port number to which this datagram is destined. The receiving host will use this port number to de-multiplex the SCTP datagram to the correct receiving endpoint/application.

Verification Tag

The receiver of this 32-bit datagram uses the Verification tag to identify the association. On transmit, the value of this Verification tag must be set to the value of the Initiate tag received from the peer endpoint during the association initialization. For datagrams carrying the INIT chunk, the transmitter sets the Verification tag to all 0's. If the receiver receives a datagram with an all-zeros Verification tag field, it checks the Chunk ID immediately following the common header. If the chunk type is not INIT or SHUTDOWN ACK, the receiver drops the datagram. For datagrams carrying the SHUTDOWN-ACK chunk, the transmitter sets the Verification tag to the Initiate tag received from the peer endpoint during the association initialization, if known. Otherwise the Verification tag is set to all 0's.

Adler 32 Checksum

This field contains an Adler-32 checksum on this SCTP datagram.

To understand the SIGTRAN protocol suite, it is important to know the lineage of SCTP. The protocol was directly motivated by the need to transport telecommunication-signaling messages over an IP-based network. The signaling system SS7 has been the dominant bearer of control information in telecommunication networks where high performance and reliability are critical to existing services and applications. However, the SS7 signaling network is logically a separate network that requires dedicated network infrastructure and only shares some physical resources with regular user traffic. Stream Control Transmission Protocol is primarily defined as a new IP transport protocol existing at the same level as UDP and TCP. In this approach, signaling messages are exchanged over a common packet-switched (IP-based) network instead of a logically separate network but, it is instructive to note, this new protocol has been shaped largely by the fact that telephony signaling has rigid timing and reliability requirements often prescribed by government regulations.

CHAPTER 11

QUALITY OF SERVICE (QoS)—
OVERCOMING ONE OF THE BIGGEST
CHALLENGES

In the early days of VoIP, ISP or Enterprise facilities were enhanced to accept digitized voice, split it into packets, and present it to the Internet connection as data. Correspondingly, packets received from the remote PC would be reassembled into a stream of digital voice and sent to the PC's speakers.

While it was feasible to hold a conversation in such a fashion, there are some fundamental problems with this architecture.

- The voice quality was terrible. The Internet connections had no pre-allocated bandwidth and were subject to variable packet delay. These contributed to jittery conversations, with chunks of speech missing.
- There were no standards that defined how the voice was packetized and how connections were established and managed. Subscribers had to use the same VoIP package in order to talk.
- Only simple (phone) network topologies were supported—point-to-point connections between known IP addresses. There were no defined interfaces between the Internet and the PSTN.
- Only simple call services were supported—basically, conversations. None of the services normally associated with the PSTN (such as call forwarding) were supported.

These limitations made it impossible to deploy VoIP on the scale required to make it usable by businesses or a large-scale home-user base.

The next stage in the evolution of VoIP was to define an architecture that would support integration between the PSTN and IP networks. This would provide a signaling capability for call management as well as defined media paths through the IP network (with reserved bandwidth for real-time media).

Several groups began work in this area, some cooperatively, others in isolation. These included

- ETSI—the TIPHON project
- ITU—H.323 working groups
- The Softswitch Consortium
- The IETF—MGCP working group, among others.

The work undertaken by these groups established a common architecture that defines the interfaces between the PSTN and IP networks to support voice (or any other streamed media) over IP. Some new network node types have been introduced, and their responsibilities and functions are also defined.

Providing QoS in a circuit-switched infrastructure is relatively easy as technology is already available. However, providing QoS in IP infrastructure is a major challenge. Although there are several approaches to the problem, there are two main QoS models that can be considered for deployment: differentiated services (DiffServ) and integrated services (IntServ). Although still a topic for research and scrutiny at the Internet Engineering Task Force (IETF) and other research communities, these models are, by and large, well accepted. Service providers and large enterprises use the DiffServ model because it scales well. However, each one approaches the problem in a very different way.

The primary goal when the Internet Protocols were designed was to provide an effective technique for interconnecting existing networks. Other important goals were survivability in the face of failure and generality in supporting various services and applications. To reach these goals, the IP protocol suite was designed to provide a connectionless datagram network that does not require signaling and per-flow forwarding state in network elements. It has turned out that the architecture scales to large networks and supports applications making many end-to-end connections (e.g., the World Wide Web). One design trade-off made to enable interconnection was to support only best-effort service at the network level and rely on endpoint functionality to obtain various levels of service. Best-effort service provides adequate support for traditional data applications that can tolerate delay, loss, and varying throughput along the path. However, in networks carrying high loads of traffic, this type of service is often inadequate for meeting the demands of applications that are more sensitive to packet loss and delay (e.g., telephony, video on demand, multimedia conferencing, etc.).

Traditionally, demanding real-time applications have been built on networks that are vertically optimized for the particular application. This design principle results in networks that are efficient for their purpose, but do not easily support new applications and are in many cases incapable of efficiently multiplexing applications with varying resource demands. It has turned out that the cost of running several different networks in parallel is high.

IP was from the beginning designed to be a general communication solution. IP technology is now recognized to be cost effective and suitable for supporting both traditional data applications and delay-sensitive real-time applications. To provide expected service for real-time applications, logically (and physically) separate IP networks are used. Each IP network serves only a subset of sensitive applications (e.g., IP telephony) with quite predictable resource requirements. By limiting the range of applications, the total resource demand can be predicted, so that the network can be dimensioned using the same traffic models as are used for vertically optimized networks. The benefit of cheap IP equipment is obtained without requiring support for dynamic service provisioning in the IP technology. Network operators now aim at cutting the overhead cost of maintaining several parallel networks. One current trend is to simplify the infrastructure by running all kinds of applications, with various network service demands, in the same logical IP network (i.e., the Internet). This means that the application heterogeneity in IP networks is increasing.

In the research and standardization bodies, the development of QoS support has progressed from providing signaled solutions for the Internet (somewhat resembling the solutions used in vertical networks) to now recognizing that more stateless solutions are favorable. RSVP (Resource Reservation Protocol) is the signaling protocol standardized to set up per-flow quality of service in routers supporting IntServ (Integrated Services) along the path. Each router along the path performs admission control and then recognizes the individual application data streams to provide the service expected. It is argued that this model is too complex and does not scale enough to be used in the backbone of the Internet. Others argue that the model scales well enough to be used close to the edges of the network.

The scalability problems of per-flow QoS management in routers have resulted in a new approach being taken in the IETF, known as the differentiated services (DiffServ) architecture. The objective is to provide scalable QoS support by avoiding per-flow state in routers. The basic idea is that IP packet headers include a small label (known as the DiffServ field) that identifies the treatment (per-hop behavior) that packets should be given by the routers. Consequently, core routers are configured with a few forwarding classes and the labels are used to map packets into these classes. The architecture relies on packet markers and policing functions at the boundaries of the network.

One advantage of differentiated services is that the model preserves the favorable properties that made the Internet successful; it supports scalable and stateless forwarding over interconnected physical networks of various kinds. The standard model is, however, limited to differentiated forwarding in routers and therefore the challenge lies in providing predictable services to end users. Qualitative services (relatively better than best-effort services, but depending on where the traffic is sent and on the load incurred by others at the time) can be provided by relying only on DiffServ support in routers and resource management mechanisms for semi-static admission control and service provisioning. To provide quantitative (minimum expectation) service, resources must be dynamically administrated by the resource management mechanisms and involve dynamic admission control to make sure that there are sufficient resources in the network to provide the services committed. There are specific requirements for resource management mechanisms. To provide service to end users, they must detect network resources and schedule them for the committed service at any granularity (e.g., for a port range, for aggregate traffic between a pair of subnets, etc.).

In core domains, dynamic aggregated resource management (e.g., per destination domain, per port range for IP telephony, etc.) must be provided for scalability reasons. ISPs need support for negotiating bulk bandwidth with each other by using advance reservations and time-dependent contracts (e.g., time of day, day of week, etc.). In enterprise networks, there are often well-provisioned LANs and bottleneck leased lines to interconnect sites. Enterprises need tools for deploying and controlling new internal services (e.g., multimedia conferencing) that require certain amounts of resources.

The QoS-based Network Infrastructure

Quality of Service, or QoS, is usually handled at the router or network edge level, although some IP-PBXs can also set it. Each edge device handles QoS differently, but for the most part it comes down to assigning a specific type of network traffic as a priority.

By giving VoIP traffic more priority than, say, HTTP Web traffic, you can guarantee that if network bandwidth becomes scarce, VoIP traffic will still move through the network where HTTP traffic may not. This means that HTTP packets will be dropped in favor of VoIP packets. Providing QoS for VoIP takes on new meaning as you move your enterprise over to IP-based telephony. Your phone system has to work every time, and you—and your callers—can't be bothered with poor voice quality because of network traffic. Monitoring tools such as Shomiti's Surveyor are also critical to logging and monitoring your overall QoS.

Good quality voice over IP (VoIP) conversations depend on maintaining strict constraints for packet loss, delay, and jitter. (Contrast this with traditional data network traffic, where the focus is on response time or throughput.) To deploy a VoIP with QoS service on a converged network, it is always recommend to check all the following matters:

Use the G.711 codec end-to-end, unless lack of capacity requires compression.
Codecs are the hardware or software used to convert from analog to digital and back. The G.711 codec gives the best voice quality, because it does no compression, introduces the least delay, and is less sensitive than other codecs to packet loss. Other codecs, such as G.729 and G.723, consume less bandwidth by doing compression, but this introduces delay and makes the voice quality very sensitive to lost packets.

Avoid bursts of consecutive lost packets.
Packet loss occurs because of congestion or electromagnetic noise. It can also occur when jitter is high and the jitter buffer is too small to compensate. Increased bandwidth and good tuning can often reduce network congestion, which, in turn, reduces jitter and packet loss.

Use a small speech frame size and reduce the number of speech frames per packet.
When voice traffic is a stream of small packets, the effect of one being lost is less severe than losing a big packet with multiple speech frames inside it. A good target is 20 ms of speech per frame, with one frame per packet. Of course, using small packets increases the total bandwidth requirement, because each packet requires its own fixed-size header.

Always use packet-loss concealment.
Packet-loss concealment masks the loss of a packet or two by using information from the last good packet. Packet loss can occur randomly or in bursts. PLC helps with random packet loss. The cost for doing PLC is minimal, because it's usually already part of the codec processing.

Actively minimize one-way delay, keeping it below 150 ms.
One-way delay = propagation delay + transport delay + packetization delay + jitter buffer delay. Voice quality degrades quickly when the total one-way delay is greater than 150 ms. Propagation delay is the time to travel the physical distance from end to end. For example, it may take a signal about 100 ms to go from Dallas to Singapore. When the traffic has to cover long distances like this, make sure the network path is as direct as possible. Transport delay is the total time spent inside each of the devices in the network, such as switches, routers, gateways, traffic shapers, and firewalls. Some devices add more latency than others; for example, a software firewall running on a slow PC adds more delay than a dedicated hardware-based firewall. Look at the number of hops traveled by the voice traffic. Reduce the number of hops and find ways to reduce the latency in the devices that are the worst offenders. Packetization delay is the fixed time needed for the codec to do its job. The G.711 codec imposes the smallest packetization delay. In contrast, the codecs that do compression add delay ranging from 25 ms to 67 ms. Also, avoid converting from one codec to another along the network path. Jitter buffer delay is used to dampen variations in packet arrival rates. If the network delay is low and the jitter is high, you can afford to have a larger jitter buffer than in a network where the delay is already high.

Avoid using slow speed links.

If you're considering VoIP, don't consider using it extensively on slow serial links. Upgrade the bandwidth on those paths so the VoIP traffic and existing data traffic have plenty of room to breathe.

Use RTP header compression for slow-speed links.

VoIP traffic uses the Real-time Transport Protocol (RTP) to encapsulate the speech frames. RTP header compression (called "cRTP") can reduce the 40-byte RTP headers to a tenth of their original size, lowering the bandwidth consumed between routers. Enable it when there's a link on the path with throughput less than 500 Kbps. So why not always use cRTP? It adds latency.

Use data packet fragmentation for slow-speed links.

Routers use data packet fragmentation to cut large packets into smaller ones, and then reassemble them at the other end. On slow links, this helps assure that small VoIP packets don't get delayed behind large data packets. Enable it for link speeds below T-1.

Use call admission to protect against too many concurrent calls.

Testing can help determine the maximum number of concurrent VoIP conversations that your network can carry with good quality. For example, your results may show that 50 calls can be carried well. You would set the call admission threshold in the VoIP server or gateway to prevent the 51st simultaneous call from being established.

Use priority scheduling for voice traffic.

Voice traffic has stricter packet-loss, delay, and jitter requirements than traditional network traffic, so it makes sense that it should receive an appropriate quality of service (QoS). A preferred QoS method is to mark VoIP packets with the DiffServ setting for Expedited Flow (EF). Also, consider using Weighted Fair Queuing (WFQ), which raises the priority of low-volume traffic. Giving VoIP higher priority helps routers decide which traffic to forward first when congestion occurs. Watch for adverse affects on the existing data traffic, though.

Upgrade your data network for VoIP and converged network.

Most data networks today aren't ready to carry good-quality voice conversations. It's easy to assess whether a network is ready or not, though, because VoIP traffic can be simulated and its characteristics can be measured and analyzed. This means you can make all the changes needed in the network and assure success before beginning an expensive VoIP deployment.

Open-Standard VoIP Solutions

Although not every converged communication solution on the market adheres to open standards, most vendors are beginning to recognize the appeal that a standards-based approach offers their customers. Legacy PBX solutions were typically proprietary. This meant that the original vendor was the only company that could develop enhanced services and new applications for the system. Not only did this approach limit the number of possible applications that might be available for the enterprise customer, but it also generally required specialized telecom technicians to reconfigure or add functionality to the PBX. With a standards-based architecture, the enterprise customer can select services from multiple providers and products from various vendors. Similar to router networks today (which are based on Internet standards), these components will work seamlessly together. Additionally, enterprises can selectively add service components and applications as they are needed, and these additions will easily integrate with the core platform. The standard that is becoming the cornerstone for communication convergence is Session Initiation Protocol (SIP)1. Previously, VoIP vendors relied primarily on the H.323 protocol. Essentially, H.323 is a telephony-centric standard modified for IP. It was initially developed to support ISDN videoconferencing and, as such, retains many of the development limitations of the traditional telephony world. SIP, however, was developed by the Internet Engineering Task Force (IETF) and is based on HTTP, so it is a Web-centric protocol designed to enable real-time communication applications over the Internet. This allows SIP to more effectively take advantage of Internet architecture.

Here are some of the key advantages of SIP.

- SIP is network indifferent; SIP-based applications can be delivered via frame relay, ATM, or IP backbones.
- Because SIP is an Internet protocol, it facilitates the integration of communication applications with other Web-based applications.
- Using SIP, features can be introduced and supported at the endpoints (similar to PC clients). This reinforces the IP Communications model whereby applications can be easily added to the network without being intrinsically tied to a single, underlying vendor.
  Customer A Premises is a "greenfield" site. Here, native IP phones convert voice signals into packets before sending them to the router for transmission over the data network. When using IP phones, no PBX system is required, because these features (voice mail, conferencing, etc.) are hosted in the WorldCom network. Network gateways are used to convert traditional voice signals to data packets and vice versa, which enables the enterprise data voice network to interoperate with the PSTN. Customer B Premises has an existing PBX and phones in place. For this enterprise, a gateway router at the customer's premise turns voice calls into data packets for transport over their data network.

By combining voice with IP, converged networks offer many appealing application possibilities. Converged solutions derive control, addressing, protocols, security, and other mechanisms from paradigms established within IP networks and on the Web. Thus, in addition to traditional PBX calling features, a whole new set of services and applications becomes possible that were extremely difficult or impossible to provide in traditional, telephony-centric environments. Imagine what can be enabled when multimedia communications such as voice, Web content, e-mail, video, and instant messaging are easily integrated. Developing applications in an IP environment empowers a larger pool of Web developers to create communication applications. When combined with an adherence to open standards, the converged approach to communication solutions will help to spur innovation in application development. Enterprises will benefit from the availability of a multitude of communication applications. Even personalized solutions can be quickly created and rapidly deployed. In the traditional PBX model, one would be hard pressed to find a vendor that would spend the time and money to incorporate a feature that only appealed to a single customer.

The following list describes some of the enhanced features that are emerging:

- Unified Messaging: The ability to consolidate and synchronize all messaging, including voice mail, e-mail, and fax.
- Presence: The ability to determine who is available, where they are located (home, office, or traveling), and what type of communication is available to them (phone, instant messaging, video, or e-mail).
- Mobility: The ability for an end user to access their personal communication services (incoming voice and text chat sessions, e-mail, voice mail, etc.) with any device (phones, PC/laptop, PDA) and on any network, including WLAN networks within the enterprise as well as wireless "hot spots" in hotels, airport lounges, and convention centers.
- Caller Preference: The ability for an end user to specify which calls to accept based on who is calling, the time of the call, and what devices are available. For instance, a user could specify that calls from their boss be sent through to their cell phone, while all other calls are sent to voice mail during the meeting.

With an IP-based voice platform, enterprises can deploy IP phones. These phones support traditional PBX functionalities (call forwarding, camp, hold, caller ID, etc.), but will also be able to integrate with the Internet and Internet-based services (central directory information, Web pushes, e-mail integration, etc.). Companies that sell IP phones today include Avaya, Nortel, Siemens, NEC, Mitel, and Cisco. Most of these phones follow the standards-based approach for convergence. This means a company can choose between phone manufacturers and know whichever phone they choose will interoperate with the company's underlying network. Also, applications created by different vendors will work seamlessly with the IP phone handsets, just as third-party applications can integrate with PDAs today. After all, the network recognizes the IP phone as just another IP end device.

One feature that IP phones enable that is not possible in the traditional PBX world is mobility. By plugging into any Internet connection, IP phones can be used at home (or anywhere in the world) and work the same as in the office. IP phones will enable communication mobility in a manner analogous to logging on to an ISP account (MSN©, AOL©) with a laptop; no matter where you log on, the service is able to identify you and deliver your account-specific personalized information (e-mail, instant messages, home page configuration, etc.).

Converged networks enable numerous administrative efficiencies for the enterprise. In addition to the end-user benefits of mobility, there are administrative aspects relating to adds, moves, and changes to the PBX system that become much easier in an IP environment. For instance, legacy PBX systems require an adjunct to be installed at each remote branch, and it generally requires dispatching a technician to install or rewire phone-switch configurations whenever a change is made to the platform. With a converged solution, IP phones bring "plug-n-play" ease to network reconfiguration. When a new phone or entire branch office system comes online, it will automatically notify the network of its resources, capabilities, and dialing plan information. These new stations will integrate with corporate databases and receive all relevant information from the other network systems, thus reducing administrative overhead. Scaling the system is also easier because the architecture does not rely on hardware telephony cards with physically defined port-capacity limitations. Theoretically, the network can recognize and support as many nodes as the enterprise wishes to add to the network (and these nodes can be located anywhere in the world that has Internet connectivity). Converged networks also offer enterprises greater flexibility with respect to determining which applications and services to manage internally and which to outsource.

Reasons for Degrading Quality

The influence of the converged network on the perceived quality of the IP telephony call is significant. The major sources of degradation are delay, jitter, and packet loss.

Delay

Because voice calls are real-time, full-duplex communications, end-to-end delay of packets can have severe repercussions on usability of the VoIP solution. Delay of less than 150 ms is considered acceptable, while delay of more than 400 ms is considered to be unusable. The delay experienced in a call occurs on the transmitting side, in the network, and on the receiving side. Most delay on the transmitting side is due to codec (packetization and look-ahead) and processing delay. In the network, most delay stems from transmission time (serialization and propagation) and router queuing time. Finally, the jitter buffer depth, processing, and, in some implementations, polling intervals add to the delay on the receiving side.

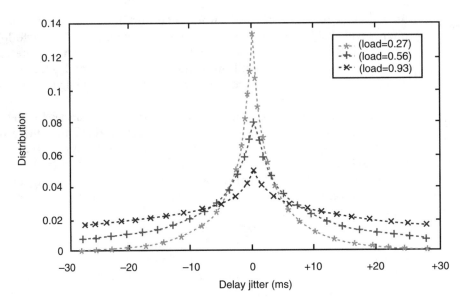

Figure 11.1

Sample picture of Delay Jitter at different traffic loads.

The delay experienced in a packet network is classified into the following types: the Accumulation Delay, the Packetization Delay, and the Network Delay. Each of these adds to the overall delay experienced by a user. The accumulation delay is caused by a need to collect a frame of voice samples for processing by the voice coder. This delay depends on the sample time and the type of voice coder used.

The accumulated voice samples are next encoded into a packet, which leads to the packetization delay. Finally, once this packet is sent through the network, it

Parameter	Fixed delay	Variable delay
CODEC (G.729)	25 mSec	
Packetization	Included in CODEC	
Queuing delay		Depends on uplink. In the order of a few mSec.
Network delay	50 mSec	Depends on network load.
Jitter buffer	50 mSec	
Total	125 mSec	

Table 11.1

Sample delay budget table

experiences some delay to reach the destination. This is caused by multiple factors, including the processing done by each intermediate node in the network to forward the voice packet, the capacity of the underlying physical medium, and so on. Delay in transporting a voice packet through the network leads to two main problems, namely Echo and Talker Overlap. Echo becomes a major problem when the roundtrip delay through the network becomes greater than 50 milliseconds. To avoid the problems of echo in calls, echo cancellers are required. Because packet networks introduce higher end-to-end delay, and hence have a greater roundtrip time, echo cancellation is an essential requirement for a voice over packet network. Apart from this, in cases where the end-to-end delay becomes greater than 200 msec, the problem of Talker Overlap surfaces. This is experienced by one talker overlapping the other talker, and can be extremely annoying. Also, a delay of more than 200 msec feels like a half-duplex connection and cannot be claimed to be an interactive session network. Of these, one scheme proposes to play the last packet received again, to compensate for the delay in packets. However, this scheme cannot work if there is a burst of lost packets. Another scheme proposes to send redundant information, increasing the bandwidth requirements. Thus, lost packet compensation is an essential requirement for supporting VoIP.

Jitter

Jitter occurs because packets have varying transmission times. It is caused by different queuing times in the routers and by possible different routing paths. Jitter results in unequal time spacing between the arriving packets, and requires a jitter buffer to ensure smooth, continuous playback of the voice stream. In IP networks, not all packets suffer the same amount of delay. Variations in packet delay, also known as jitter, cause VoIP packets to arrive at their destination in uneven patterns. This can result in degraded voice quality. Typically, the solution to jitter problems is to increase the size of the jitter buffer in VoIP components. However, this solution increases overall delay and must take into consideration network delay characteristics.

Figure 11.2 shows a sample of waveform distortion that can cause time-base distortion.

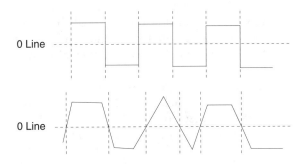

Figure 11.2

Waveform distortion.

The top waveform represents a theoretically perfect digital signal. Its value is 101010, occurring at equal slices of time, represented by the equally spaced dashed vertical lines. When the first waveform passes through long cables of incorrect impedance, or when a source impedance is incorrectly matched at the load, the square wave can become rounded, fast risetimes become slow, and reflections in the cable can cause misinterpretation of the actual zero crossing point of the waveform. The second waveform shows some of the ways the first might change; depending on the severity of the mismatch you might see a triangle wave, a square wave with ringing, or simply rounded edges. Note that the new transitions (measured at the Zero Line) in the second waveform occur at unequal slices of time. Even so, the numeric interpretation of the second waveform is still 101010! There would have to be very severe waveform distortion for the value of the new waveform to be misinterpreted, which usually shows up as audible errors—clicks or tics in the sound. If you hear tics, then you really have something to worry about.

Packet Loss

Most packet loss occurs in the routers, either due to high router load or to high link load. In both cases, packets in the queues might be dropped. Packet loss also occurs when there is a breakdown in the transmission links. The result is data link layer error and the incomplete packet is dropped. Configuration errors and collisions may also result in packet loss. In non-real-time applications, packet loss is solved at the transfer control protocol (TCP) layer by retransmission. For telephony, this is not a viable solution because retransmitted packets would arrive too late and be of no use.

Packets can be easily lost because of network outages, network re-routing, congested routers (deliberately dropping packets), or a congested device, such as an Internet Protocol (IP) telephone or a Public Switched Telephone Network (PSTN) gateway. Packet loss can range from 0 to 20 percent or more depending on the type of network, number of hops, and amount of congestion. IP data can be reliably transmitted using the Transmission Control Protocol/Internet Protocol (TCP/IP), which makes an initial handshake between devices before sending to ensure connectivity, and automatically resends packets if they are lost. Human conversation cannot wait for re-sent packets, nor afford the time to do handshaking. Therefore, speech is sent using the Universal Datagram Protocol/ Internet Protocol (UDP/IP), which does not re-send information nor require an initial interaction with the receive end. The result is relatively fast transmission but no guarantee of delivery.

IP packets can also be delayed if they take a longer route or spend time in a device queue. Buffering can be used so that devices wait for late packets. However, speech packets that arrive too late, beyond the buffer length, must be discarded in order to meet timing requirements. One of the best solutions to late and lost packets is to engineer the network to minimize the situations that cause delays and packet loss. However, in spite of excellent engineering, some packet losses can remain during periods of congestion and a truly robust system requires send- and/or receive-based concealment solutions.

LAN-based Prioritization—IEEE 802.1p/Q and IP Precedence

For QoS, network devices must have a way to classify traffic. One approach is for the application or end system to signal the desired level of service to the network by tagging packets at Layer 2 or Layer 3 like 802.1p/Q or through a signaling protocol (such as RSVP) that allows flows to be associated with different priorities. These approaches have the disadvantages of requiring compatible QoS functionality in both end systems and network devices and of being prone to "mistrusted" end systems that could request highest priority regardless of application. A far simpler approach to QoS in the enterprise network is for the network device to classify traffic by recognizing application flows through inspection of the packet headers at Layers 4–7 of the OSI Model.

IEEE 802.1p adds 16 bits to the Layer 2 header, including 3 bits that can be used to classify priority (the tag). Frames with 802.1p implementation are called "tagged frames." The standard specifies 6 different priorities, which do not offer extensive policy-based service levels. In addition, implementing 802.1p in a network with non-802.1p switches could lead to instability, because older switches would misinterpret the unexpected 16 bits specified by the standard. Implementing 802.1p in older networks could require a costly upgrade of all switches. Most importantly, because the Layer 2 header is only read at the switch level, the boundary routers, where the bottlenecks occur, cannot take advantage of prioritization based on 802.1p unless it is mapped to a Layer 3 prioritization scheme. While prioritization is achieved within the switched network, it is lost at the LAN/WAN boundary routers.

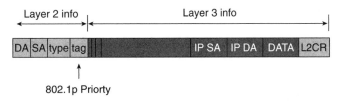

Figure 11.3

A sample of an 802.1P header.

IEEE 802.1Q standard is a protocol that provides standardized VLAN and prioritization capabilities to Ethernet networks through four new bytes of data in the header. Although this changes the fundamental design of the frame's header, it also allows end-station equipment to continue using mathematics to figure out the frame's contents, without breaking infrastructure equipment at the same time. With this design, 802.1Q frames appear to contain data for ethertype "8100" networks, so if an 802.1Q-compliant end-station sees an ethertype of 8100, it can pretty much assume that the frame is formatted for 802.1Q data (and then conduct additional tests to verify the assumption). But older hubs and repeaters won't be affected, because the frame still looks like Ethernet.

The downside of this design is that older end-stations can't see whatever data is inside the 802.1Q frame. Thus, print servers won't see print jobs submitted by 802.1Q devices, database clients won't find servers with 802.1Q adapters, and a host of other problems will arise from mixing and matching old and new gear together without careful planning. To keep these problems at a minimum, most of the 802.1Q switch vendors will be providing support for legacy equipment on a per-port basis, stripping the excess header data from 802.1Q traffic that is being sent to a legacy device. Although the 802.1p/Q mechanism cannot operate on an end-to-end basis in an internetwork, it does provide a relatively simple method of defining and signaling an end system's requirements within a LAN environment, which assists QoS on converged network.

IP Precedence provides the ability to classify network packets at Layer 3, analogous to the functionality of the 802.1P protocol at Layer 2. With IP Precedence configured, network packets traverse IP Precedence devices according to the priority you set. Priority traffic is always serviced before traditional traffic. The first attempt to use this field was defined in IETF RFC1349. IP Precedence is considered the best effort Quality of Service at layer 3. Packets are prioritized and then queued accordingly to traverse the network. IP Precedence was designed in IPv4 by the IETF. It uses 3 bits of the 8-bit Type of Service (ToS) field of an IP header. There are 8 classes of services in IP Precedence. The classification range is 0–7, where 0 (zero) is the lowest and 7 is the highest priority.

IP Precedence is a scheme for allocating resources in the network based on the relative importance of different traffic flows. The IP specification defines specific values to be used in this field for various types of traffic.

The basic mechanisms for precedence processing in a router are preferential resource allocation, including both precedence-ordered queue service and precedence-based congestion control, and selection of Link Layer priority features. The router also selects the IP precedence for routing, management, and control traffic it originates.

Precedence-ordered queue service includes but is not limited to the queue for the forwarding process and queues for outgoing links. It is intended that a router supporting precedence should also use the precedence indication at whatever points in its processing are concerned with allocation of finite resources, such as packet buffers or Link Layer connections. The set of such points is implementation dependent.

Although the Precedence field was originally provided for use in Department of Defense systems where large traffic surges or major damage to the network are viewed as inherent threats, it has useful applications for many non-military IP networks. Although the traffic-handling capacity of networks has grown greatly in recent years, the traffic-generating ability of the users has also grown, and network overload conditions still occur at times. Because IP-based routing and management protocols have become more critical to the successful operation of the Internet, overloads present two additional risks to the network.

- High delays may result in routing protocol packets being lost. This may cause the routing protocol to falsely deduce a topology change and propagate this false information to other routers. Not only can this cause routes to oscillate, but an extra processing burden may be placed on other routers.
- High delays may interfere with the use of network management tools to analyze and perhaps correct or relieve the problem in the network that caused the overload condition to occur.

Implementation and appropriate use of the Precedence mechanism alleviates both of these problems. Most of the new IP QoS tool kit in routers and switches is the IP Precedence bits. Devices at the edge of the network may classify traffic as deserving a certain Class of Service using these bits. Core devices can then use the bits to provide differing types of service to different flavors of traffic.

Differentiated Services

Multimedia applications and convergence changed the situation—usage and users must now be distinguishable and network processing needs to be tailored to each class of traffic. For example, a real-time telephone call should not be handled the same way as a large file transfer if they share a link. Differentiating among QoS requirements can be done by classifying each packet and using this information in the switching decisions. DiffServ can be used to aggregate traffic into a limited number of classes of service on an end-to-end basis because it operates at Layer 3.

The DiffServ objective is to provide scalable QoS support by avoiding per-flow state in routers. The basic idea is that IP packet headers include a small label (known as the DiffServ field) that identifies the treatment (per-hop behavior) that packets should be given by the routers. Consequently, core routers are configured with a few forwarding classes and the labels are used to map packets into these classes. The architecture relies on packet markers and policing functions at the boundaries of the network to ensure that the intended services are provided.

DiffServ, unlike earlier IETF efforts on QoS for IP, takes a very simple, practical, and scalable approach. The strategy of DiffServ is to employ a range of standard Per Hop Behaviors (PHBs) at each router, where each PHB effectively determines the QoS treatment a packet receives at that router hop. This standard describes how to mark IP packets so that they receive a particular PHB by using what is called the Differentiated Services Code Point (DSCP), a six-bit field in the IP packet header that allows for 64 possible PHBs. It also describes the characteristics of the standard PHBs (and their associated DSCPs) and which DSCPs may be used for customized local or experimental use within a DiffServ Domain. A DiffServ Domain is a logical IP network where PHB definitions are applied consistently across router hops.

One standard PHB is Expedited Forwarding (EF). EF is designed for implementing a low-latency, low-loss, and/or low-jitter service class, such as that desired for high-quality VoIP conversations. Or it can be used simply for virtual leased line capability. EF can be implemented with a strict priority over other service classes to achieve low jitter, though other approaches are allowed by the standard that may yield somewhat different service characteristics. Next is a family of twelve PHBs called Assured Forwarding (AF), which is organized into four sets of three each. AF's four service classes each offer low packet loss for data applications that demand something better than Best Effort but do not need EF treatment. There are also eight DSCPs reserved for a set of PHBs known as the Class Selector PHBs. Some of these PHBs may be used for network management or control traffic. Among these eight is the default Best Effort PHB. It is inferior to both AF and EF, and the other Class Selector PHBs. Best Effort is used in the event that no other PHB is specified for the traffic. It may also serve as a PHB for demoted traffic (i.e., traffic that arrives in excess of metering profiles). The standard, however, only requires two of the eight Class Selector PHBs to have forwarding behaviors that are actually different from each other. It is also worth noting that neither AF nor EF are required for a router to be DiffServ Compliant, although without these, there is little hope for multi-vendor or multi-service provider compatibility.

Defining service classes, supported by a DiffServ domain, is not in the agenda of the DiffServ working group. DiffServ domain administrators need to agree on some bilateral terms to develop end-to-end IP services. These bilateral terms are based on Service Level Agreements (SLAs). DiffServ SLAs are domain limited, giving network operators flexibility to define services to best suit their customers' needs and to optimize the utilization of network resources. At the same time, network operators are limited by this very flexibility in that the services defined must map closely to those defined by adjacent operators. It is therefore necessary to develop an end-to-end QoS framework. A general consensus among IP network operators is critical while defining the service classes. A safe rule of thumb is to classify the applications into three major categories: real time, non-real time, and best effort.

Two DiffServ example services are Premium service and Elastic service, which supplement the existing Best Effort (BE) service. These services are constructed over the PHBs provided by the DiffServ framework. Note that service names are trivial and implementers can name them differently. For example, Premium, Elastic, and Best Effort might just as well be named Platinum, Gold, and Silver.

1. Premium service: Designed for real-time interactive applications, Premium service emulates traditional leased-line service. It is based on ingress–egress pair-based SLAs. An example of such service is IP telephony. Premium Service employs EF PHB.
2. Elastic service: Designed to cover a wide range of the non-real-time applications from video-on-demand to BBE (Better than BE) traffic such as important web access and ftp. Elastic service employs AF PHB group.
3. Best Effort service: This is the traditional default service that already exists and may also include all packets that could not be classified into one of the preceding two classes. Best Effort service will not be developed in the paper.

DiffServ provides a powerful and practical means of applying QoS to distinct IP applications. As a result it is rapidly gaining widespread acceptance. Most leading routing vendors and service providers are implementing at least a few distinct forwarding behaviors (typically two to four) in order to provide some form of premium services (i.e., anything better than Best Effort).

QoS Reservations—RSVP

The Resource Reservation Protocol (RSVP) provides a means for the sender to signal a request for QoS along the path to the receiver. RSVP defines a "flow specification" that includes the desired transmission rate and/or desired bounds on transit delay. If all network components are able to support the request, then the receiver asks for the required resources to be reserved and the sender is informed that the flow can be initiated. RSVP is receiver initiated in order to support multicast operations and to allow the requirements in each direction of flow to be different. It should be noted that RSVP provides only the signaling function and cannot, by itself, ensure that the node's resources actually get reserved. RSVP was originally designed to be a means of setting up higher-quality "connections" within the Internet. Unfortunately, the overhead involved in maintaining a connection for each flow limits the scalability of RSVP. More recently, RSVP has been re-directed for use in local signaling within a DiffServ environment.

A host uses RSVP to request specific qualities of service from the network for particular application data streams. A reservation is made along this flow by every node supporting RSVP. Reservations are made for simplex flows, i.e., in only one direction, but reservations can be made for many-to-many multicast sessions. One reservation corresponds to a unique *session*.

It is important to notice that RSVP does not enforce the reservation it makes. It is like a booking system: it checks whether a host can request a given service and, if so, it records the reservation and forwards the query, but the node (IP router, ATM switch, etc.) still has to enforce the given reservation.

IP packets are routed independently from one another, which means that two consecutive packets can use totally different paths to reach their destination. RSVP is not a routing protocol, thus it has the same constraints: It can make a reservation on a certain path, and then packets will use another path on which nothing has been reserved, leaving the reserved path unused. That is why RSVP establishes *soft states*: Reservations need to be periodically refreshed, and if this does not happen (e.g., packets use another path to reach their destination because of a broken link), the state times out and the reservation is deleted. This mechanism permits dynamic changes of reservations.

RSVP operates on top of IP (IPv4 or IPv6), occupying the place of a transport protocol in the protocol stack (but it does not transport any application data). The RSVP traffic is UDP encapsulated, which means that it is unreliable. The packets sent use UDP port number 46.

A *session* is defined as a data flow with a particular destination and transport-layer protocol. Each session can be either a unicast or a multicast session. The reservations are described by a *flow descriptor*, which is composed of a *flowspec*, specifying the requested quality of service and describing the data flow, and a *filter spec*, which enables the packets to be filtered. The receiver of a data flow is made responsible for the reservation, in order to handle dynamic group membership easily: The sender sends an RSVP message to a host saying that it is willing to send information to it, and then the receiver of the message requests a reservation.

The RSVP process does two basic operations, as follows:

1. Receive and process a message
    The request is analyzed. If the changes requested are possible, the state is updated.

2. Forward the request
   The message can be modified by the RSVP process.

An RSVP implementation has two decision modules. *Admission control* determines whether the node can supply the requested quality of service and *policy control* is responsible for checking administrative permissions for a particular reservation.

It is noticeable that the RSVP process needs to know where to forward the message. As RSVP is not a routing protocol, it has to consult the local routing tables to get this information. Therefore an implementation needs an interface with the routing mechanisms of the node.

There are seven different RSVP messages shown in Table 11.2.

Message type	Description
Path	RSVP process initiation: set up path state
Resv	Reservation request
Path Teardown	Deletion of *path state*
Resv Teardown	Deletion of a *reservation state*
Path Error	Sent to the sender of a Path message that created an error
Resv Error	Sent to the sender of a Resv message that created an error
Confirmation	Confirm that a reservation has been done correctly

Table 11.2

The two most important messages are the Path message and the Resv message. Two hosts willing to request a service from the network proceed as follows:

1. The sender (of the data flow) sends a Path message to the receiver. All along the path, every node supporting RSVP records this Path message and forwards it.
2. When the receiver receives the Path message, if it accepts the communication, it sends back a Resv message following the reverse path the Path message used, requesting the service for the session. Each node records this request, makes the corresponding reservation, and forwards the message upstream. As RSVP has a "soft" state; the sender and the receiver periodically send Path and Resv messages respectively to refresh the state. If those refreshes do not occur, the reservation is automatically deleted. The refresh period is calculated in such a way that a reservation will not be deleted because a UDP packet transporting an RSVP message got lost. States can also be deleted explicitly downstream using a Path Teardown message, or

upstream with a Resv Teardown message. If an error occurs during the reservation process, the originator of the message is warned by an error message. As with other messages, every node supporting RSVP will do the necessary modifications in its tables if it gets an error message. If a Path message created the error, then a Path Error message is sent. A Resv Error will be generated in the case of an error due to a Resv message. Finally, a host can ask to receive an acknowledgement once the reservation is made. A Confirmation message is used to inform it that the reservation was successful.

Policy-Based Network for QoS

QoS Policy-Based Network and Systems Management refers to the capability of Networks and Computing Systems, using policies that reflect business goals, in an automated manner, to control access to Networking and Computing resources. Policy-Based Systems are able to determine, based on load measurements (for example, cpu utilization), network congestion, restriction, and prioritization of access, and allowance of free access. As a result of the deployment of emerging real-time and mission-critical applications, the traditional "best effort" IP network service model is unsuitable. There are several QoS solutions offered by equipment vendors, service providers, and software vendors that are targeted toward the enterprise customer.

Most organizations already discover the elements in their infrastructure automatically. The policy perspective assesses the ability of infrastructure elements to support a policy system. For example, can a network device prioritize traffic as it forwards it (supporting resource assignment and forwarding policies)? If so, it can be an active part of the policy system (an enforcer); if not, it cannot play a role unless it is upgraded.

Planners use this information to evaluate the policy capabilities of their current infrastructure; some devices or management services may require upgrades or replacement before they are part of a policy-based system.

- Prioritizing traffic
- Bandwidth reservation
- Traffic classification
- Admission control
- Traffic shaping
- Load balancing
- Alarm generation

The goals of the policy-based network architecture include the creation of a standards-based system that addresses both the enforcement and the administration of policies. Specifically, the system will perform various functions including the following:

- *Policy Administrator* Creating a distributed, hierarchical system to coordinate and manage policies between policy management devices, directories, and various policy enforcement devices using a standards-based architecture approach.
- *Interface to Directory Database.* Provides access to higher-level, user, and application-level policy information via data stored in common directory repositories. This high-level database information can then be translated into actual policy enforcement actions.
- *QoS Gateway.* Providing end-to-end policy enforcement and management via standards-based signal provisioning protocols including Differentiated Services, ToS, RSVP, MPLS, and 802.1P
- *Bandwidth Broker.* Providing bandwidth brokerage services allowing an automatic system for multiple domains to negotiate service-level guarantees.
- *Centralized Monitoring and Accounting.* Provides centralized policy-based accounting and remote monitoring services.

In a standard policy-based network, policies consist of two components.

- A set of *conditions* under which the policy applies. This might include parameters such as user name, addresses, protocols, and applications types.
- A set of *actions* that apply as a consequence of satisfying (or not satisfying) the conditions, including bandwidth guarantees, access control, service load-balancing, cache redirection, and intelligent routing.

These conditions and actions consist of a series of passive and active components on the network. As a simple model, this would include a *policy manager*, which is the central policy administration and directory repository point, and a *policy enforcer*, which consists of remote active management components that make up the local policy decision and enforcement points throughout the local wide-area networks.

The Policy Enforcer can be a simple router that makes policy decisions based on a field in a particular "tagged" packet. Alternatively, the Policy Enforcer may be a piece of equipment that locally consolidates and analyzes traffic flows and network conditions in order to perform complex network actions such as

- *Traffic Conditioning and Shaping.* This includes things such as traffic prioritization, traffic guarantees, and bandwidth management.
- *Policing.* This includes access control, user authentication, and remote login.
- *Tagging/Signal Provisioning.* This includes translating and relaying signal provisioning information (RSVP, Differentiated Services, 802.1P, MPLS) through the network.

- *Server Resource Control.* This includes advanced enforcement capabilities such as server load balancing and cache redirection.

Typically, this more complex piece of equipment will sit at the edge or border of a network domain and relay, via signaling protocols, overall bandwidth requirements to the internal network equipment. When referring to Policy Enforcement, this document refers primarily to these edge devices.

The Policy Management functions are responsible for coordinating administrator input with other external and internal policy information and translating this information into concise network terminology. The Policy Management service is responsible for coordinating the flow of bandwidth between various devices that make up the input and output to specific domains. By managing the critical border locations in a specific domain and then having the enforcement devices communicate, via standard signaling protocols to the internal devices, this system creates a complete end-to-end policy-based network solution. Policy Manager functionality can be distributed and hierarchical in structure, representing various physical and logical domains for management of specific Policy Enforcement devices. As an example, an enterprise domain can be defined for the whole corporation that includes general policies and management structure that affect every user, application, or resource. Individual departments "inherit" the higher-level corporate policies but can then define their own enforcement policies for the clients and servers that they control. The corporate network administrator can delegate administration rights to the local department's administrator.

The Policy Manager consists of various logical pieces that can each run on separate servers or combine into a single system, including the *Policy Editor*, the *Interface to the Directory Database*, and the *Policy Decision Administrator:*

The *Policy Editor* provides the network manager with capabilities to centrally configure rule-based policies that take high-level concepts and translate them to control of services within the network infrastructure. The Policy Editor can be accessed directly by network managers for input via a Java/web interface or can be part of a centralized management structure provided by systems management products such as HP OpenView or Tivoli. The editor will, in general, store information on LDAP-based database directory. Communication with the directory can be through the Policy Administrator or through third-party directory applications. The Policy Editor will have added capabilities for translating generic policy information stored in the directory to actual policy rules.

The Directory Database Server provides the central repository for policy information. In general, the Directory Database resides on a Directory Server as a third-party LDAP-based application provided via companies delivering directory-related products such as Microsoft, Netscape, IBM, and Novell. The Directory Server is responsible for the storage of a wide range of information including e-commerce, user login, and network yellow pages as well as network-specific policy information. The directory can be grouped in a distributed and hierarchical fashion for sharing data. The key to the directory is the wealth of applications that allow users to organize and correlate the wealth of information that subsequently gets stored into the database. A user can, for example, enter their entire customer, list into the directory and then assign specific administrative and policy information on each customer, including network-specific bandwidth policies.

The Policy Decision Administrator acts as a central command center for policies in the network. The Policy Decision Administrator is software based and can run simultaneously on the directory server or as separate entities that communicate with the directory via LDAP. In some cases, parts of the administration functions may reside on one or more of the Policy Enforcers, allowing the enforcer to run independent of external devices.

Policy Enforcement can come about by performing a specific service at the Policy Enforcer. Alternatively, simple policy devices, such as core routers, can make policy enforcement decisions based on information that is "relayed" through the network by various standards-based "tagging" or signaling protocols.

End-to-End signaling mechanisms

-RSVP.

Differentiated Services

LAN-based signaling mechanisms

802.1P/Q

A Policy Enforcer can work as a QoS gateway whereby it will translate and enforce end-to-end policies to each specific domain.

A policy network with QoS is a set of rules that describe the actions that a system takes in different circumstances. These actions are triggered automatically by the appropriate policies and they support an intelligent and adaptive environment for delivering business-critical services. Building an effective policy system is best done with a phased approach. Although the phases and their steps are discussed separately, in reality there will be some overlaps. The first is a reality check, gathering a realistic assessment of the current capabilities of the network infrastructure, followed by policy creation, policy deployment, and evaluation and tuning. A reality check is the foundation for making the appropriate decisions for constructing an effective policy system. Organizations need a robust policy system if they are to meet the challenges of managing dynamic business-to-business and customer-to-business applications. Supporting mission-critical applications, delivering QoS, and maintaining security are among the hardest challenges IT professionals face today.

CHAPTER 12

DEPLOYMENT STRATEGIES—ISSUES AND ENTERPRISE CHALLENGES OF DEPLOYMENT

In this chapter we will primarily be discussing the deployment of a Voice over IP solution. These techniques will also apply to other converged network applications such as video on an IP network. We will discuss the process to determine the right solution, ensuring the reliability and manageability of the solution, and give some tips on how best to deploy the solution.

Criteria of Selecting Vendors and Products

Step One: Evaluate Needs

The first step in deploying a converged network is to determine exactly what your organization's needs are. These needs are primarily driven by the applications the end users will deploy and how these users intend to access and use these applications.

Before looking into how to deploy a converged application, the network managers need to determine the application needs of the end users. These needs will vary widely from one user type to the next. Some key needs that need to be identified include

Size: The first thing that needs to be determined is how big the required network is. This is not just how many telephones there are, but how many of what types of users there are. Another item that needs to be considered is how much growth is expected in the network.

End-user Voice Features: The needs of all of the user types need to be clearly known, from the needs of the executives all the way to the needs of the back-office staff. Determining which features are needed is usually the most critical determining factor as to what solution works best for you. A critical mistake many network managers make is to underestimate the end-user features used in their current environment.

Many times it is believed that only basic features such as conference, transfer, forward, or hold are used, while, in reality, more advanced features are in use such as call bridging, intrusion, paging, and executive/assistant support features. The main problem encountered here is these advanced features are only used by very few users, but these tend to be very important users.

Specialized User Needs: Many organizations have users who have unique, specialized needs. The users include those who may work remotely or at home either all the time or part time. Other users may be traveling or otherwise off-site but need access to voice features from the main location. There are also other users who may need specialized access such as TDD (telephone device for the deaf).

Specialized Applications: There are often organizations within an enterprise that need specialized voice applications. One of the most common ones is the contact center. Sales groups, customer service groups, and help desks often deploy specialized voice functions such as most-idle agent queuing, skill-based routing, and other contact-center-specific functions. These users often need access to specialized services such as Call Management reporting, call recording, and back-office records to help in the support of their customers.

Messaging: Most enterprise business voice users need some sort of access to messaging. This messaging can include voice mail, unified voicemail/email systems, or just access to someone to answer the call and take a message. When deploying a voice system it is important to determine the user's needs not just for the messaging system itself, but also for coverage or forwarding of calls to the message system and notification that a message is waiting.

Applications Access: Many end users need or want access to special applications for their voice calling needs. These applications include items such as access to corporate directories to facilitate speed dialing, access to contacts data bases, or access to computer-based applications to support voice calls.

Network Access Needs: Almost all voice users will need to access and be accessed from the outside world. An important need to be considered is what services are required from the network. Services can include items such as direct inward dialing (DID), caller ID, network access features like Centrex, or call-re-route. Also, the amount of outgoing and incoming calling needs to be supported and it should be determined whether it is more cost effective to use analog or digital facilities.

Multi-site Support: Many enterprises have multiple locations. Another factor that needs to be considered is the interaction between these sites. Needs such as site-to-site direct access, uniform dialing, and feature transparency should also be determined.

The best way to determine these user needs is to perform a detailed survey of all of the user types in a given enterprise. This survey can give the network management team information on exactly what features and services will be required of the new network.

Determining the exact needs of your particular organization will be a key determining factor in deciding what solution and what vendor works best for you.

Step Two: Determine Skills Fit

Another item that needs to be evaluated before choosing a solution and a vendor is the skill sets that your organization has. It doesn't help anyone if an ideal solution is chosen, but no one knows how to implement, support, and use it.

In the past, voice networks and data networks were completely separate. This separation did not just apply to the physical networks, but, in most cases, the personnel who supported voice and data as well. Voice networks and data networks have typically needed very different skill sets to deploy and support.

Supporting converged networks requires a mixture of voice and data support skills. Some vendors design their converged network offerings to look more like data networks. Some look more like voice networks. Understanding your organization's skill sets can help determine which vendor's solution may match best with your team.

Also, by understanding your implementation and support personnel's strengths and weaknesses, you can better determine what vendor may be able to complement and enhance those skills. This knowledge will also determine what skills need to be augmented before deploying a converged network application.

Step Three: Choose Solution

Once you have determined your needs from an application, user, and support perspective, then it is time to evaluate and chose a particular vendor's solution.

The factors most used to choose a solution are scalability, price, ease of use, and voice features.

Scalability

Scalability means different things to different people. To some it means the ability of a solution to grow from a stand-alone branch-office installation today to an enterprise-wide converged network in the future. To others it may mean the ability to link together many small sites.

In small to medium locations, the right solution is usually determined by the feature needs of the users.

In large, single-site organizations, large, feature-rich solutions may be the best bet as they will not be outgrown in a short period of time. These types of organization may be best served by the larger, PBX-based solutions on the market.

To companies in a more distributed environment, scalability may have nothing to do with how many phone lines a single chassis will support. Many times small, simple solutions let companies start small, with a limited upfront investment, and grow from there. They can also be rolled out across the network of a large company that has many small locations. Oftentimes in these types of environments, the simpler IP-based server can work the best.

Price

When deploying a converged network application like VoIP, "price" can mean much more than the cost of acquiring and implementing a given solution. Other factors, such as cost of support, ongoing operation, and infrastructure, are just as important in determining an overall price of a converged network. Saving on capital investment and operational costs is one of the primary drivers of convergence, but the relative efficiencies different products can offer depend partly on your existing infrastructure.

The cost of a given solution is often determined by the ongoing support needs of the system. For example, some IP-router-based VoIP architectures can have low starting costs but need extensive, high-priced expertise to deploy and support. The same is often true of the extensive voice services on a PBX-based system. These voice services need specially trained personnel to exploit effectively.

Another major cost factor is the existing environment. If a large PBX system is already in place that can be upgraded to VoIP support, that may be a lower-cost solution than starting from scratch. On the other hand, if the data network is not voice ready the changes to the network may offset that cost if the VoIP support can be built as part of the network upgrades.

Again, understanding the needs of the users, the capabilities of the support personnel, and the capabilities of the network are crucial in the determining the best solution.

Ease of use

Ease of use, for most people, is defined by what they already know how to use. For example, an administrator that has been programming features on a PBX will say a PBX-based VoIP solution is the easiest to use. An IP network engineer will probably say a router-based VoIP solution is the easiest to use. Most organizations that deploy VoIP find that in-house expertise tips the scale in favor of a particular vendor.

Ease of use does not just apply to the administration but also to the end users. Having all of the user features in the world does no good if the users can't make them work. Again, it is very important to know the skills and comfort level of the administrators and users in your environment to help determine what solution works best for you.

Features and functions

Most of the time the PBX-based VoIP vendors do the best job with regard to voice features. However, most organizations use only a fraction of a PBX's features. You need to identify this subset and use it as a checklist when evaluating voice over IP systems. If your users have to sacrifice voice features they've been employing, it won't be well received. Pay particular attention to key applications and functions such as voice mail and direct extension dialing among sites. Here, too, the PBX vendors tend to have an edge.

If you are an IT-intensive company that tends to customize products, you might want to map your in-house programming expertise to the various voice over IP alternatives. Those organizations that have programmers in house who have a core competency with a given architecture such as Microsoft or Cisco can get added value out of the customizability of these systems.

No matter which type of voice over IP vendor you choose, you'll have to put off deployment if you plan to hold out for every feature you want. Most VoIP architectures are still under development. It is important to determine what solution meets your needs today as well as what solutions will evolve to meet your needs in the future.

Other considerations

One other consideration that needs to be made before choosing a particular solution is to evaluate the vendor of that solution. Oftentimes a solution may seem perfect, but the vendor cannot deliver the solution or the vendor ceases to support a solution or the vendor may cease to exist at all.

There are a few questions that should be asked before choosing a vendor.

- Does the vendor have the necessary expertise in both voice services and IP networks to deploy the solution?
- Does the vendor have the resources or partners to design, deploy, and support the solution?
- Is this product part of the vendor's ongoing product line?
- What is the financial viability of the vendor and its partners?

That last question may be the most important one. In the current economic environment, companies are experiencing hard times and corporations that seemed very large and solid in the past may or may not survive into the future. It pays to do a lot of research into a vendor before committing your future network to their products.

Implementing Deployment Schedules

Once the solution and vendor are selected, the next phase is to plan and implement the solution. There are several steps needed to deploy a converged network. These steps take the converged network from initial needs assessment to final deployment.

Step 1: Application Needs Assessment

As we discussed in the previous section, the first step in the implementation process is to fully assess the needs for the voice network. These needs include voice user needs and also specialized needs such as contact centers, voice messaging, and inter-site networking.

Step 2: Network Audit

The next step in designing the converged network is to take stock of what currently exists in the data-only and voice-only networks. Review the existing equipment and evaluate its capabilities and operating costs. Determine existing facility costs and whether the current networks meet planned voice and data needs. Identify upcoming multimedia applications and new services that are needed to keep the company competitive, and determine their impact on the network as much as possible.

Determine the service quality of both voice and data to the user community, and which areas need improvement. Perhaps a new Web-based financial application does not have the user response time required to make employees productive. A traffic study may be necessary to look at current voice and data traffic patterns, as well as to determine which applications and protocols are using the data network today and what their bandwidth needs are. Perhaps bandwidth on some links in the network can be lowered, while others need to be increased.

Business policies should be reviewed or established regarding what traffic types (applications) on the network should get priority, which applications are mission-critical to the business, what bandwidth will be allocated to each, and what traffic on the network should be disallowed. The result of this analysis is at the basis of bandwidth provisioning as well as the Quality of Service (QoS) techniques that will be deployed in the network to ensure certain traffic types, including voice, meet the needs of the end user community and support business goals.

Step 3: Setting Network Objectives

Once the current traffic and network use baseline is established, the next step is to set objectives for the integrated network. First, determine the dominant traffic types the integrated network is expected to support, and the service-level expectations and agreements (even if informal) with user communities within the organization. Also, consider how closely voice and data functionality will be tied together. These appraisals will help in the selection of the appropriate technologies.

Setting voice quality objectives will establish your organization's acceptable delay and compression limits.

Determine the voice traffic load the network can absorb and still meet the baseline data networking requirements. Determine the policies (priority and bandwidth) regarding mission-critical traffic types on the network.

There is usually not a single reason for migrating to a converged infrastructure, but some reasons are more important than others to a particular business. Being clear about these reasons will ensure meeting business goals when technologies are selected and network design decisions are made. Determine the top and lower priority reasons for migration in your organization. These may include

- *Deploying new applications*—to enhance business efficiency and/or employee productivity
- *Cost savings*—on communications facilities, equipment purchase and maintenance, bandwidth efficiency, and long-distance or international voice call expenses
- *Increasing competitiveness*—including e-business interworking with suppliers, partners, and customers, offering new services and products, and improving customer care
- *Consolidating network management*—including simplifying equipment vendor relationships, leveraging cross-team skills, consolidating network management platforms and applications

Determine the technical requirements of the converged network. Is any-to-any calling important, or only remote office to headquarters? Will fax traffic be carried? Should the dialing plan currently in place be retained or can it change? Will the converged network only carry On-net calls, or also be leveraged for On-net to Off-net (and vice versa) calling? Must the PBX configuration and software level remain unchanged, or will these be upgraded at the same time? Clearly defining these objectives will guide technology and network design choices.

Step 4: Technical Design and Capacity Planning

In this step the network designer needs to determine the design criteria that need to be added to the data network to support the VoIP application.

There are several key VoIP network parameters that need to be set up to support the user and network needs determined as outlined. These parameters include

Voice Coding and Compression—Which voice coding and compression schemes are used is determined by the needs of the user application and the available network bandwidth. For example, a contact center environment may need the best voice quality possible, while internal calls across a WAN can use a compressed encoding scheme to control bandwidth usage.

Quality of Service Techniques—Once the needs of the users are known along with the capabilities of the IP infrastructure, a QoS technique or techniques can be chosen. QoS, again, is the means whereby applications such as voice or video that need higher priority transport through the network can be supported. This is usually done by either labeling and prioritizing voice or video packets or by reserving bandwidth in the network for these services.

At this stage in the deployment it is recommended that policies for all traffic types on the network be defined. That way as new applications are added there is already a framework in the network plan to support prioritization in the network.

Dialing Plan Design
In a converged network the voice gateway must understand the dialed digits in order to set up a voice call to the appropriate destination. In an IP network this amounts to translating the dialed digits to an IP address. To aid the voice gateway in this mapping, a dialing plan is implemented on the voice gateways in the network, similar to those on PBXs and in the PSTN.

Several factors must be considered regarding the dialing plan to ensure a smooth cutover and a manageable network. First, must the dialing plan from the user's perspective remain unchanged? Or will a new PBX "access code" be introduced to allow users to choose the IP network or the PSTN for their calls? Second, will the IP network only carry On-net calls, or also provide gateways to the PSTN? For On-net calls, only the company's private dialing plan needs to be implemented. If Off-net calls are also accommodated, the gateway may have to translate an internally dialed number to a publicly accessible number the PSTN will understand, including adding/deleting country and regional area or city codes.

Digit manipulation can be used to manipulate numbers such that the end user interface can be what the business desires, and digits can be changed to adapt to country specifics or public vs. private network access. Translation tools exist to manipulate both the calling and called digits of a call, which is often important for billing or tracking purposes.

Capacity Planning—Based on the proposed network design and the required number of trunks between locations, the required bandwidth can be calculated. Bandwidth calculations should take into account compression, overhead, and utilization. Each of these will vary, depending on which transport technology is chosen. Bandwidth efficiency techniques such as RTP Header Compression (cRTP, applicable only to VoIP) and VAD (Voice Activity Detection) can be deployed. VAD, or Silence Suppression, keeps packets from being sent when there is no live voice present. For example, a conversation has a speech path in both directions at all times, but typically only one person is speaking. VAD will suppress the listening party's "empty" packets from being sent onto the network. In addition, on a multiservice network, voice bandwidth is only needed when calls are actually being made. The benefit is that when calls are not being made, perhaps at night when the office is closed, the bandwidth is available for data applications such as nightly backups. It is important to design the network such that voice calls that will oversubscribe the allocated network bandwidth are kept off the network and rerouted via an alternate path such as the PSTN. This is another voice QoS category called Call Admission Control. For example, if the WAN access link from a branch office is provisioned to carry no more than five simultaneous calls, the sixth call must not be allowed onto the network as it will impair voice quality.

Step 5: Network Roll-out

In a network of any realistic size, this does not happen overnight, and current applications usually cannot be disrupted while the new network is cut over.

Many customers prefer to start with a proof of concept laboratory test to prove that the decisions on vendor products and network parameters will give satisfactory service. Experience is gained with the new technology and configurations independent of the live network and the design can be proven or adjusted.

Next a small pilot test phase may be rolled out to a select number of users who have been prepared or trained to report on the voice quality they receive and perhaps have an alternate way of dialing a call if the network does not behave the way they need to conduct business. Many configuration and feature adjustments may happen during this pilot phase to tune the network, services, and design.

When a successful pilot phase has been completed, a roll-out plan for the remaining sites can be put in place and each can be cut over at a time that is convenient for the support group. Existing equipment may have to be upgraded, either software or hardware or both, to be voice-ready before the cutover of that site can take place. The logistics of the cutover phase depend on the user community, the geographic challenges of getting to a site, the potential impact of disrupting service to a particular application or site, and many other factors.

Ensuring Transparency

Transparency for the end users is highly desirable, unless the goal is to deliberately change their environment, providing dramatically new features and capabilities. Most leading solutions claim to have implemented the commonly used features, and it would seem that transparency is a non-issue. However, the availability of full-feature functionality should not be undervalued.

Taking features away from users will almost guarantee unhappiness. Meeting even 95 percent of every user's current requirements means 0 percent satisfied users. Understanding the specific needs of the users and evaluating the proposed system capabilities against them is critical to avoid significant support and re-training costs. In the world of IP voice, particular attention should be given to support of existing terminals, fax machines, and conferencing services.

Another area in which care must be taken to ensure transparency to the user is during a phased implementation. At some point in the implementation there will be users on both the new and the old system at the same time. One of the biggest challenges is to support the user's needs and support business during this transition period.

One of the arguments often made for using PBX- or KTS-based VoIP architectures is that this transition period is taken care of because the voice server supports both the old and new environments. This is usually a true statement, however, there may be user or network reasons to transition to a new, pure VoIP solution. In that case, care must be taken to ensure the continuity of service and as much end-user transparency as possible during the transition.

In Figure 12.1 we show a hypothetical network before the transition to a converged VoIP application.

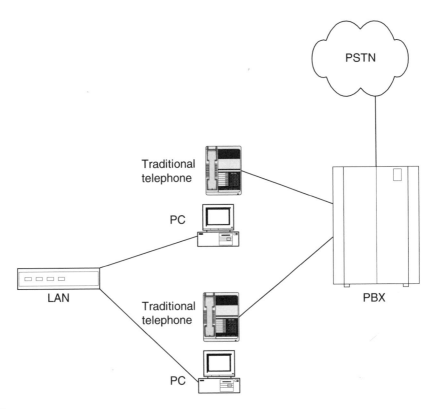

Figure 12.1

Network before deploying convergence.

Once the converged network has been planned, designed, and piloted, this customer has chosen to migrate to the converged architecture in phases. Figure 12.2 shows the network during the transition to the new architecture. It is in this phase that care must be taken to ensure transparency between the users on the old system and those on the new.

In many cases, as illustrated, some form of temporary link needs to be established between the two networks. Many standard interfaces are available to do this, such as tie trunks. To keep transparency to the end users digital tie trunks can be deployed along with protocols such as ISDN-PRI and Q-Sig, which give some feature transparency between the two servers.

Care must also be taken so that shared user services, most especially voice messaging, can be accessed from both servers. This may mean supporting either a link that allows for central voice mail sharing, like Q-SIG, or temporarily connecting the voice mail service to both networks.

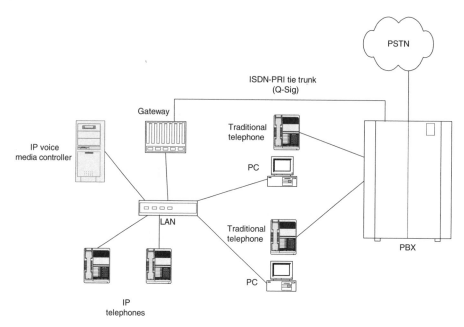

Figure 12.2

Partial transition to converged network.

Once all users migrate to the new architecture, these temporary links can be removed (see Figure 12.3).

These temporary connections need to be planned in advance of the network implementation and often add to the cost of the final solution, but they are vital in ensuring a smooth transition for the end users.

Availability of the Overall Solution

One of the single most daunting issues facing converged networks is the incredible robustness and reliability of the current public voice networks. Users expect voice services to work all of the time, flawlessly. Users will expect this type of reliability from voice services on the converged network as well.

There is a quote voice network vendors like to use. No one is entirely sure where the quote came from, but it is a good example of users' attitudes toward the reliability expectations in a voice network.

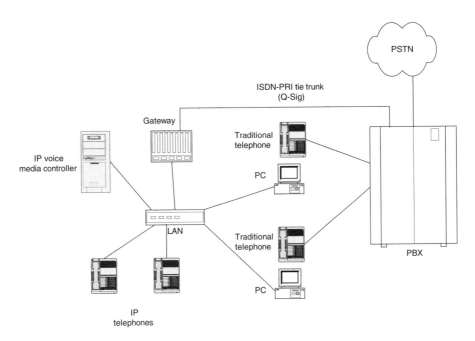

Figure 12.3

After migration to converged network.

"Data availability is an absolute necessity, Dial Tone is a God-given right"

The quote is, at best, politically incorrect, but it makes a point as to the expected availability of the voice network.

The high availability of voice networks is so expected that most people use it as the final point of emergency response. Access to public safety services such 911 is usually done over the telephone. Enterprises that deploy converged networks need to be sure that this level of reliability is maintained. This is an important consideration in light of safely and liability concerns.

Availability (up time) of the overall solution should be evaluated carefully. Not only are the specifications and architecture of the vendor's products important, but also the inherent availability of the IP network they use. A single network is good, but it is also a single point of failure.

Reliability of networks is usually measured by something known as "9s". Networks are said to have "4-9s" or "5-9s" reliability. The "9s" refer to the percent availability of the service. "4-9s" means 99.99 percent available, "5-9s" is 99.999 percent available. To put this into real terms, the Table 12.1 gives the equivalent of "x-9s" of reliability in downtime per year.

x–9 s	Percentage	Downtime/year
2–9 s	99	3 days, 15 hours, 36 minutes
3–9 s	99.9	8 hours, 45 minutes, 36 seconds
4–9 s	99.99	52 minutes, 33.6 seconds
5–9 s	99.999	5 minutes 13.56 seconds

Table 12.1

This means that a "5-9s reliable network" is only down for just over 5 minutes a year (including, in theory, all of the time needed ot do upgrades and preventative maintenance)!

So in designing a converged network care must be taken to minimize possible sources of down time. This usually means that the IP network must be designed with a high level of fault tolerance and redundancy. This is often the single largest cost factor in deploying a converged VoIP solution. Care must be taken in designing a converged network to provide the availability that is wanted in a reasonably affordable architecture. This is an area where the network designer must clearly know the needs and tolerances of a user before deciding the best solution.

Even with a fully fault-tolerant and redundant IP infrastructure, it is difficult to give the reliability of legacy voice networks simply because there are more intermediate devices such as IP switches and routers between the voice server and the user. Often a choice needs to be made between offering new applications and technologies and keeping the reliability of the old services. Many real-world converged networks end up being hybrids of both converged and non-converged users (converged users typically being those who need the new applications, non-converged users the ones who need high reliability above all).

Data Network Management

Data network management takes on new importance, given that the data network is becoming the infrastructure for all of the enterprise's critical communications. Competent staff backed up with solid network management systems capable of monitoring the critical multi-media network parameters (latency and jitter, for example) will be core requirements in a converged environment. Strong vendor support is also important here, complemented with remote diagnostic and management capabilities.

Network management tools, if properly deployed, can provide the network administrator with a complete view into any enterprise network. With the advent of converged networks, it is imperative to have network management systems enable the following capabilities, at a minimum:

- Network discovery and topology maps
- Inventory control and configuration management of networked nodes
- Report generation, system logging, and analysis of the respective data.

Simple Network Management Protocol (SNMP) enables network management applications to retrieve data from the various converged network components in a standardized fashion. SNMP can be used to support voice services from the network management tool.

Voice services also have unique systems management needs that must be a part of the network management tools. Some of these unique tools are items such as Call Detail Recording, for call billing and call path tracing for troubleshooting needs. The management tool also needs to be able to administer end user features and access to the PSTN.

Because of the needs to support certain levels of performance for voice calls, the network management tool also needs to be able to give parameters on the network that support voice performance. Item such as latency, jitter, and packet loss for the voice packets need to be monitored to ensure good voice quality and to give the network administrator a tool for fixing problems with these parameters.

Quality of Service (QoS)

Throughout this book we have constantly stated that the most important aspect that determines the success or failure of a VoIP implementation is the Quality of Service (QoS) given to the voice (or video) services on the converged network.

Even in a well-managed IP network with adequate bandwidth and good latency characteristics, it is necessary to implement a QoS strategy to prioritize traffic and ensure that voice, data, and video requirements can be consistently met. This may require the replacement of some legacy data network elements with newer elements that implement the appropriate QoS standards (DiffServ, RSVP, 802.1p, 802.1q, etc.).

One of the hot new development topics in the industry today is the deployment of overall QoS schemes for a network and tools that are specifically designed to support and manage these QoS schemes. The generic term for this type of network QoS enforcement scheme is a Policy-Based Network. Several vendors have introduced the concept of policy-based network management. Most of them work using a server, known as a Policy Server or Policy Manager. In the policy server the network administrator sets up policies as to how certain data items should be treated in the network.

Network policies include items such as the priority levels for the various applications and protocols in the network. Most policy servers allow the administrator to set up certain priority levels for items such as management protocols, voice services, email, internet access, and data services access. The more sophisticated policy servers even allow those priorities to be varied for different users or at different times of the day.

Once the network policies are set, the policy server then translates these policies into control messages to the various network components to set up the QoS parameters that will support these processes. For example, the policy server will tell the IP telephones and gateways to give their voice packets a certain 802.1Q and/or DSCP tag and then tell the network switches and routers what level of priority to give to packets tagged in that way.

The main benefit of a policy server is to set up network-wide QoS deployments from a central site. This helps avoid the need to implement QoS parameters on every network device and endpoint separately. This not only saves effort, but helps eliminate error and allows for consistent performance across the entire network.

Security Needs

Security is another prominent issue in the deployment of the converged network. A converged network administrator must worry about both data network and voice network security issues along with new issues that arise based on the unique characteristics of the converged network. Converged network security is a topic that would require an entire additional book to cover completely, so we will just try to cover the main points briefly.

Voice network security issues predominantly revolve around toll fraud and public safety issues. Preventing toll fraud usually involves keeping unauthorized users from accessing outside lines and using them for their own benefit. This is usually prevented on two fronts. First, system features such as classes of service and facility restrictions prevent the unauthorized user from accessing the outside lines. Second, the systems administration tool needs to be protected from anyone gaining access to change those restrictions. Deployment of passwords and firewalls to prevent unauthorized access to these management services is also needed on the network.

Public safely issues involve both access to services like 911 and also the ability to prevent or at least report and track malicious calls into the system. Data Network security also involves preventing access to systems from unauthorized parties as well. Preventing systems "hacking" is a well documented art in itself. Data networks are also vulnerable to other issues such as denial-of-service attacks and viruses.

Converged networks are vulnerable to all of these issues and more. In a converged network there are new worries like viruses disrupting voice servers and endpoints as well as denial-of-service attacks on voice services and terminals. Also in a converged network there are issues with the security of the voice calls themselves. Many tools exist for an entity on the network to collect and monitor packets on the network. If those packets happen to be carrying voice calls, that entity could decode and listen in on the conversation.

IP telephony vendors and customers recommend these steps to manage the security of voice over a data network.

- Separate IP PBXs on the LAN by putting the devices in different domains from other servers.
- Isolate voice traffic onto a virtual LAN.
- Limit administration access to IP PBXs among IT staff, allowing only a few to have access to the core operating system on a VoIP server.
- Limit the types of protocols that can touch the IP PBX or IP telephony network when possible.
- Encrypt voice traffic where possible. Do not send IP voice over an unmanaged or public network.

Many tools to provide security are available and should be strongly considered. Tools such as firewalls to prevent unauthorized access from outside the network, encryption to code the voice or other sensitive packets so that listeners cannot decipher the packet contents, and virtual private networks to secure connections through public media like the internet or a wireless connection are available to help secure the converged network.

Security is possibly the single largest area of concern in converged networks, so great care needs to be taken to ensure that the savings and access to new features that convergence promises are not overshadowed by a loss of user and network security.

Interworking with PSTN

Until the entire world is supported over a single, converged infrastructure, any deployment of a VoIP architecture must still have some access to the Public Switched Telephone Network (PSTN). Interworking with the PSTN can be as simple as deploying a media gateway that supports standard voice network trunks like analog central office trunks (CO), direct-inward-dial trunks (DID), or digital facilities like T1 or E1. Some networks need to use more advanced access services such as ISDN-PRI and ISDN-BRI. These types of interfaces can provide more advanced features like two-way trunks and caller identification.

Fortunately, most current VoIP architectures provide support for nearly all common network interfaces as part of their core offering. Almost every vendor's VoIP suite of products includes an appropriate PSTN media gateway in some form or another.

Many VoIP architectures can even support a protocol known as Signaling System 7 (SS7). SS7 is the internal signaling protocol in the PSTN. ISDN signaling was originally intended as the end user access to SS7, but many larger corporate and service provider networks can use a direct SS7 access point. Additionally, by incorporating SS7 into their multi-tiered systems, service providers can leverage the industry's effort to develop an intelligent, SS7-based PSTN and quickly introduce the same set of advanced intelligent network (AIN) features to the IP telephony world.

In this chapter, we have discussed the deployment of the Converged IP network. We discussed the steps of selecting products, scheduling deployment, and ensuring the transparency, availability, and quality of the network.

We have especially emphasized the need to fully understand the needs of your particular deployment, including the needs of the users, the equipment already in place, and the skill-set of your users and network support staff. This understanding is the biggest key for successfully deploying a converged IP network.

CHAPTER 13

PREPARE FOR CONVERGENCE-READY NETWORK—A SOLID APPROACH

Against the backdrop of immense change now occurring in converging telecommunications and data communications industries, network service providers (NSPs) and IT and network managers today face unprecedented challenges as well as unprecedented opportunities for growth, profit, and cost savings in network expenditure. With the right decisions at the right time, regional Bell operating companies (RBOCs), competitive local exchange carriers (CLECs), service providers (SPs), Internet service providers (ISPs), and enterprises running large private networks can deploy Voice over IP (VoIP) services with great success. With the constant advancements in Voice over IP (VoIP) technologies, mission-critical voice traffic can now be carried reliably and with consistently high quality over an IP network.

Besides avoiding the tolls charged by ordinary telephone service, the integration of voice and data on a single network greatly simplifies the evolution of integrated services such as multimedia collaboration and incorporation of voice services with Web-based applications on desktops and wireless terminals. While the integration of voice and data on a single network is conceptually simple, there are numerous complex technical issues that must be understood both in the implementation of and installation of a VoIP system. For example, failing to properly configure quality of service (QoS) related parameters would result in poor voice quality and/or significant degradation of data performance.

Because IP packets carrying voice are treated just like IP packets carrying any other type of data, they are subjected to delay, loss, and retransmission. This is especially true when the network is congested. The quality of service becomes a very important issue. Losing every other word of the phone call can make the call essentially worthless. IP telephony is facing the following challenges:

- Unpredictable service quality, which relates to quality of service and reliability. Real-time applications set high requirements on the reliability and quality of service capabilities of IP networks. Protocols and techniques to ensure this must be developed. Until these techniques are widely deployed and supported by most networks, over-provisioning or proprietary methods in private IP networks remain the only way to ensure the required QoS.

- Datacom and telecom convergence-related complex system integration, Network Management Systems (NMS) integration, Customer Care and Billing (CCB) systems integration, and diversity in the marketplace. IP telephony equipment consists of new network elements that need to be integrated into the corporate and teleoperator's or service provider's network. Both physical and logical integration to the other network elements are required, as well as integration to the vital operation support systems such as maintenance, provisioning, and billing systems.

- Lack of interoperability because a single waterproof standard does not exist. There are several competing or partially overlapping standard proposals. Current IP telephony standards only ensure interoperability within a single IP telephony subnetwork. The communication between gateways or gatekeepers from different vendors remains to be standardized.

- Regulatory development will have a major impact on IP telephony. In many countries IP telephony is still unregulated but the regulatory authorities are monitoring the situation closely.

- Inertia in the legacy networks, large investments tied into legacy technologies, and people are accustomed to the old services. There is inertia in the traditional telecom services.

IP telephony requires bringing new technology into networks: logical network elements and protocols. New logical network elements are needed for call management, routing, storing call information, and so on. Signaling protocols are used to establish calls or multimedia sessions, such as multimedia conferences, voice calls, and distance learning. The IP signaling protocols are used to create connections between clients over intranets or the Internet. The main functions of signaling protocols are user location lookup, address translation, connection setup, service feature negotiation, call termination, and call participant management such as invitation of more participants. Additionally, signaling protocols are responsible for billing, security, and directory services, for instance.

One of the most vital issues for widespread use of IP telephony is to achieve international standards that enable equipment combinations of different vendors to work properly together. Currently, several organizations are developing IP telephony standards. A few important organizations are ITU-T, IETF, ETSI, iNOW!, IMTC VoIP forum, and MIT's Internet Telephony Consortium.

Two remarkable signaling protocol "standards" exist for IP telephony: ITU's H.32x series and IETF's Session Initiation Protocol (SIP). H.323 seems to have quite a good position in the current market when compared to SIP, which is a new protocol trying to get into the market. H.323 is a so-called umbrella standard for multimedia communications over local area networks, which does not provide guaranteed quality of service. H.323 belongs to the series of communications standards called H.32x., for multimedia conferencing over different types of networks including ISDN and PSTN. The H.323 specification was approved in 1996, but first drafts of the H.32x series were approved in the early 1990s. The second version, approved in January 1998, concerns conferencing over wide area networks.

SIP was developed by the Multiparty Multimedia Session Control (MMUSIC) working group of the IETF. The protocol is still under development and it is not as well known as H.323. SIP is based on hypertext markup language (HTML) and is more lightweight than H.323. SIP was originally designed for multimedia conferencing on the Internet. In addition to SIP, two other signaling protocols are considered as parts of the SIP architecture: Session Description Protocol (SDP) and Session Announcement Protocol (SAP).

Figure 13.1 illustrates a typical architecture of IP telephony network based on H.323. The network architecture consists of four types of network elements: terminals, gatekeepers, gateways, and multipoint control units (MCU). The basic configuration consists of at minimum two terminals connected to a local area network (LAN). However, in practical applications it is necessary to add some of the other elements in order to create an efficient communication system with connections to the outside world.

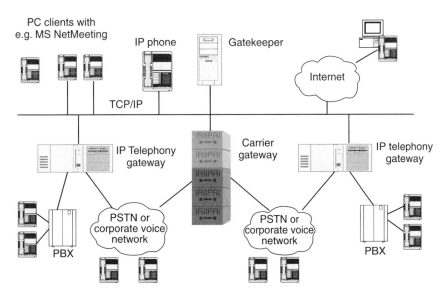

Figure 13.1

IP telephony network based on H.323.

The key challenge that the IT and network manager faces is therefore how to engineer IP networks to suit the applications and existing infrastructure. He or she has to harness the power of IP technologies to provide consistent service levels across the heterogeneous base of devices, from network switches and routers, to application and communication servers. The IT manager also has to decide on what VoIP technology to use, and make sure the VoIP network is secured so that no one can eavesdrop or snoop on any VoIP end user's conversation. It is crucial that when a IT manager chooses a vender for a VoIP application, he or she makes sure that vendor provides some type of real-time voice payload encryption mechanism (for example, RC4 encryption) that protects the end user's voice conversation.

A good rule to follow is to start with the equipment that you already have to see how you can add voice-processing quality. In this way, even if an upgrade does not work out, you still have the advantage of a familiar partner when venturing into uncharted waters. In the VoIP world, Figure 13.2 shows you ideal VoIP requirements to deploy a seamless converged network.

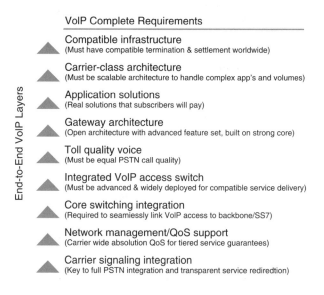

Figure 13.2

Hardware Requirements

The first step is selecting the right hardware to build out your infrastructure so it can take full advantage of VoIP technology, and support the widest range of next-generation VoIP services and applications. Make sure your choices include the following:

- Scalability
- Interoperability
- Quality of Service (QoS)
- Feature-Rich Platform, Breadth of Product
- Pervasive Platform
- Proven Reliability

Scalability

Without a fully scalable hardware platform, you'll be hard-pressed to keep up with the explosive growth in the VoIP market. Unless the platform scales easily, you may forfeit newly captured profit due to forklift upgrades. If your deployment includes PC-based solutions, it's important to make sure they are both scalable and reliable. Whether you are a growing Enterprise network, top-tier ISP or CLEC, or an RBOC concerned with market retention, the one constant you can count on in the telephony market space is rapid growth. So it's vital to invest in voice gateways, gatekeepers, and network management platforms that can scale smoothly to carrier-class capacity if necessary. There are many factors that are involved in scaling a VoIP network topology, including change in communication latency, bisection bandwidth, change in network throughput, packaging and layout cost, size and cost of expansion, change in routing and flow control, and change in performance-to-cost ratio. Many of these conventional factors can only be improved at the expense of others, thus requiring designs to either optimize for one or find an optimal compromise between many. Let us briefly discuss several somewhat atypical factors that have been the basis of a few very different proposed scalable topologies.

One of the scalability considerations in a topology refers to the number of links needed in the VoIP topology to be connected to a node as the network is scaled up. It depends on the node degree, which is the number of channels incident to a node. Link complexity has a direct correlation to hardware cost and complexity. Channels are implemented either as wide buses or optical cables for high bandwidth. Buses occupy a significant amount of space on circuit boards and suffer bandwidth limitations, while optical cables are expensive and require optical–electrical interfaces. Because integrated chips and circuit boards have limited pin bandwidth, high node degree means narrower channels, which means higher serialization latency. For scalability, a VoIP network topology should ideally have constant node degree. That is, the node degree of each node stays the same regardless of the number of times the network is scaled up.

Another tip of network scalability that an IT manager should consider is incremental scalability. It refers to the number of additional nodes that have to be added to a topology when it is scaled up in order to maintain the same topology. (Strictly speaking, incremental scalability includes link complexity, but we consider just the increase in number of nodes here.) Incremental scalability is important because it affects the ease and cost of increasing the size of a network. There might be a need for the processing power (number of nodes) of a system to be expanded but not by the amount dictated by the topology. If a converged network topology can be scaled up in small or constant increments, then the amount of additional wiring needed is probably small. The amount of modification to the routing algorithm needed, if any, is also of concern, because it might require significant software and/or hardware changes to the existing network for proper functioning, particularly for real-time applications such as voice.

Interoperability

If you are planning to support voice, data, and video in one converged network, your equipment will need to interoperate with many other voice and data gateways and switches. Although this requirement seems simple on the surface, the consequences of true interoperability are extremely far-reaching. The greater the number of installed gateways and switches the converged network can interoperate with, the more worldwide calling areas it can support. The more VoIP consortiums it can partner with, the more customers it can reach and the fewer troubleshooting problems it will encounter in the path of growth. IT managers also need to interoperate with the largest number of ISPs, POPs, and telephony standards around the world, often prior to final, formal standards ratification. Also, IT managers are strongly recommended to have a service level of agreement with service providers for QOS/COS guarantee for the VoIP network. So look for a VoIP hardware platform that supports advanced feature sets for an open and extensible architecture and, if possible, a vendor that participates in standards development.

Quality of Service (QoS)

QoS is fundamental to guaranteeing end-to-end prioritization of voice over data calls. Without a network that provides the low latency, low jitter, and required bandwidth that voice calls demand, the user experience is likely to be unsatisfactory and the service unsuccessful. The key challenge in providing high-quality voice services is control over end-to-end network delay. To provide an experience comparable to the PSTN, end-to-end delay must be less than 200 milliseconds. A review of the end-to-end components of IP-based calling reveals that in a typical application, voice compression/decompression, cross-network packet forwarding, and standard jitter buffering range between only 0 to 60 milliseconds each. By contrast, congestion delays can frequently add up to 200 ms, making network congestion the dominant factor in voice call quality. Controlling the level of network congestion is consequently the key to providing acceptable QoS for VoIP services. To meet and maintain the high-uptime requirements of voice and fax, ideally you need a comprehensive, integrated voice and data management system that offers control and support for the local network segment as well as the end-to-end services system.

Feature-Rich Platform with Breadth of Product

You need to seek out equipment that offers both depth of features and breadth of solution if you intend to minimize cost and complexity to become or remain a carrier-class player. To minimize network support costs and network complexity, all service providers or enterprise networks need a VoIP platform that can handle their full range of call volume requirements from the smallest rural POPs to the largest central offices. To meet end-user quality expectations, the VoIP gateway component of your network must support true telco-class voice features. These toll-quality voice services are achieved through several key gateway technologies, including silence detection and comfort noise generation, DTMF detection and pass-through, ANI- and PIN-based authentication, and echo cancellation and phone number/IP address translation.

One more note: When a vendor offers a true, end-to-end product solution, it's not always a sure bet that the solution includes truly integrated voice and data access switching gear. Only the integrated voice and data access platforms yield the cost effectiveness that will be profitable in the long run. This is because by integrating in a single voice/data access switch all the requirements for a complete range of data and voice call termination, the switch eliminates the need for separate data and voice networks. This in turn eliminates the space, power, spares, and management and support requirements of separate and costly voice equipment. If you're a service provider, by applying the economies inherent in voice/data switches, you can maintain a feature-rich and cost-competitive position against incumbent RBOCs. If you're an RBOC, you need to begin converging your separate networks to retain your competitive advantage.

Also important to an end-to end product offering is that it includes backbone, or core switch integration. This is critical to getting control over all segments of an end-to-end voice service. With seamless integration from network access to network egress points, you can offer the highest possible quality VoIP access service with the richest possible feature set. In carrier VoIP applications, the core-switching equipment must support millions of calls with the same call completion and robustness that people have come to expect from telephone services. Equally important, if you expect your IP voice network to evolve quickly, you need to make sure your hardware supports the advanced features in the existing Signaling System 7 (SS7) world telephone network. Services such as 800 number termination will increasingly move from the traditional circuit-switched network to the IP network. VoIP access switches and core network switches that support the most advanced SS7 features will enable you to hold onto existing markets, and enter new and valuable markets driven by the extremely lucrative but competitive cost structures of IP networks. Companies that standardize on products with less-robust SS7 capabilities will be left behind.

Pervasive Hardware Platform

Ideally, the vendor you select for VoIP hardware and end-to-end equipment should enjoy a strong market position with a popular, and consequently pervasive, product offering. In addition to support for open standards and commitment to interoperability, the sheer market strength of your vendor can let you leapfrog many hurdles that smaller niche players may face. Strong networking equipment suppliers are also likely to have deep, mature, and extensive global partnerships in which you can immediately participate and profit.

Proven Reliability

Your end users must have absolute confidence in the reliability of your network if they are to entrust it with the real-time, mission-critical voice and data applications that drive their businesses. Remember, the benchmark for voice reliability must be as robust as a central office switch. For example, your gatekeeper needs to have built-in redundancy and support overlapping call areas, to provide the necessary reliability for high-uptime voice and fax service. In sum, you need to seek out experts in remote access technology and deployment, with a long track record and an ability to leverage expertise.

Software Requirements

Be careful to shop for a solution that is as strong in terms of its OS and network management feature set as it is in its VoIP hardware offering. The following are some points to consider:

- Interoperability
- Integrated VoIP Software Platform
- Best-in-Class Applications

Interoperability

In terms of PSTN integration and global call termination, this is an ITU-H.323 world, so make sure your vendor supports the standard that defines the framework for transmission of real-time voice over IP-based packet networks. Compatibility with this evolving standard is critical not only for operation of services, but also in protecting your investment over time. Interoperability of third-party applications with your voice gateways is also necessary, because this allows for smooth network expansion and rapid, trouble-free service delivery. In addition to that, make sure the chosen VoIP vendor has a committed migration path to support future protocols and industry standards, and open platform to adapt new features and functionalities.

Integrated VoIP Software Platform

An integrated platform lets you avoid duplicate systems and administration while enabling you to bring services to market faster. A converged network needs the flexibility and choice provided by a wide range of well-integrated applications from third-party developers. Look for a vendor who can give you a choice among a wide range of "best of class" solutions for your market, so you can further differentiate your services and respond quickly to market opportunities.

Best-in-Class Applications

As a best-case scenario, your vendor should either directly supply or partner with a full range of VoIP players who offer value-added applications, such as billing and call settlement, fax, unified messaging, multimedia, wireless and voice VPN applications. This affords maximum flexibility in growing or maintaining your business and gaining profit.

Strong Support Infrastructure

A vendor that has an experienced distribution and support organization with global reach can mean the difference been a VoIP solution that hits the ground running or one that stalls as a result of unexpected challenges. Look for a network equipment and service provider who can offer a complete portfolio of service offerings from installation to ongoing maintenance that can be tailored to your exact requirements as you fine-tune your internal or external service offerings.

Infrastructure Integration

Although we have discussed the requirements for VoIP platform hardware, software, and support as separate entities, their real-world functions and impact are interdependent. For example, QoS depends on end-to-end interoperability and compatibility of both hardware and software, plus a robust feature set. Scalability involves depth and breadth of product. The pervasiveness and proven reliability of a vendor's platform affects interoperability, QoS, software integration, and the likelihood that best-of-class applications will be available for the vendor's platform. As a result, the best approach to VoIP integration is to insist on a vendor who can meet not just several, but all of the preceding requirements, then work with you to adjust them to your particular business model.

Businesses don't exist just to save money. They exist to make money, gain market share, and serve customers. That's why the most compelling aspect of converged voice/data networking may well be the new generation of applications it enables. These applications include Web-enabled call centers, unified messaging, and real-time collaboration. Take the example of a Web-enabled call center. One of the biggest obstacles that companies face in converting Web site visitors into Web site buyers is poor online interaction. In a bricks-and-mortar store, customers can ask a nearby salesperson a question that may end up determining whether or not they head for the checkout line. On a Web site, that kind of interaction is more problematic. But using VoIP, site visitors can click a button and open up a voice conversation with a real, live call center agent who can quickly address any question or problem the customer might have.

Other examples include real-time multimedia video/audio-conferencing, distance learning, and the embedding of voice links into electronic documents. In fact, the full business potential of such applications is only beginning to be discovered. But one thing is clear: These integrated voice-and-data applications will require a converged IP network. Regardless of any individual observer's opinion about when and how voice-over-data is going to explode, the fact is that it inevitably will. Companies that enter into VoIP at that point in the history of business technology without any convergence experience or momentum will be at a significant competitive disadvantage. They won't be able to gain the internal operational benefits that convergence offers. They also won't be able to effectively partner with other companies that have made convergence a core component of their technology portfolios. And they won't be able to effectively service customers who will expect VoIP capabilities. Chances are that they will wind up spending more money in an effort to "play catch up" than would have been necessary if they had started moving in the right direction earlier in the game. It is therefore essential that any company with intentions to survive and thrive in a communications-based economy take steps now to ensure its ability to compete in a future that will clearly include universal adoption of VoIP technology.

The ability to deploy powerful new integrated voice and data applications with 100 percent success is important to win over competitors and be more efficient in your business performance. So, it is important to make sure that when you deploy new technology in your network, you plan carefully to have the right resources, right vendor that can offer you the whole converged package in a seamless manner, right solutions for your business "no compromise," and back up plan for disaster recovery.

Test Converged Network

As telephone voice networks evolved to carry packet traffic, and new standards emerged to support the transport of voice over this medium, a unique set of problems arose with regard to voice quality. Voice quality had been taken for granted, as it was typically not an issue with time division multiplexing (TDM) over the public switched telephone network (PSTN). The standards that formed the backbone of the PSTN (known as SS7) provided linear wavelengths, consistent bandwidth, manageable echo, and a predictably high level of quality that delivered real-time business-quality voice anyplace in the world. The emergence of Voice over Internet Protocol (VoIP) has radically altered the world of telephony. VoIP has proven it can provide consistent high-quality voice, but it was necessary to create a tool that could cheaply and effectively measure voice quality over this type of network in order to reliably and objectively measure these results. This led to more rigorous standards by the International Telecommunications Union (ITU) for the creation of voice quality testing (VQT) methods that could be a part of Converged network testing.

Voice quality is essentially a subjective measure of the clarity, inflection, and tone of the conversation between a caller and the recipient. Although many characteristics influence the perception of quality (including environment and premise equipment, etc.), in general the human ear has come to define "acceptable" voice quality within a narrow range of values. The challenge for VQT is to model this very human expectation of "quality" with mathematical equations that predict the user's experience in terms of objective and measurable criteria. The results are testing methodologies that produce a number that corresponds to how a vast majority of users will perceive the conversation.

The concept of converged testing is straightforward at one level. It simply means integrating TDM and data side testing solutions and allowing the tester to tweak either side to understand the impact on the other. The benefits of a converged solution are fewer problems in the field, faster diagnostics in the lab, and reduced testing time and cost. Today, engineers must cobble together pieces of traditional solutions, and attempt to synthesize the results back into a coherent understanding of what the problems are and how to fix them. Converged testing solutions take that burden off of developers and test engineers and provide a step increase in ability to ensure quality of service in next-generation networks. Converged testing approaches can provide the best of both, and make it easier to create real-world conditions in test labs and to isolate and diagnose problems in systems under test.

The things that impact voice quality can be roughly categorized into three areas.

1. Delay—the time it takes for a signal to travel from the caller to the recipient

2. Echo—the reflection of the user's voice over the network and back to the user

3. Clarity—the general signal strength, fidelity, and clearness of the voice conversation

Voice Quality Testing (VQT) is an objective and comprehensive methodology for testing end-to-end network voice quality. It is important, especially with the advent of Voice over Internet Protocol (VoIP), to quantify how the average human ear would perceive a conversation, especially before implementation of the solution. Prior to modern testing methodologies, the measurement of voice quality over a TDM network was an inexact affair to say the least. Testing consisted of calculating the characteristics of various linear sound wavelengths, such as signal-to-noise ratio (SNR) and total harmonic distortion (THD). But testing was usually not necessary, as voice typically traveled over TDM without much distortion. However, Internet Protocol (IP) and Voice over IP (VoIP) changed all that. Internet Protocol is a nonlinear technology not measurable through sampling testing, and new low bit-rate voice compression codecs, such as G.729A, created packets instead of wavelengths.

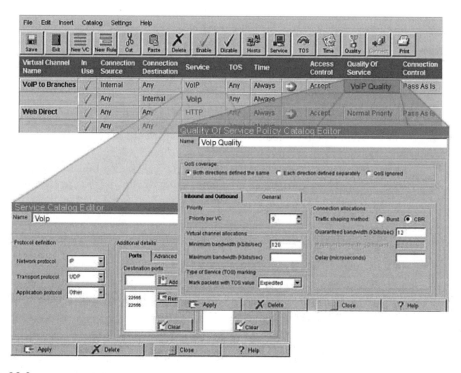

Figure 13.3
A typical snapshot of a QoS test engine.

The first step in enhancing next-generation network testing is Converged Network Emulation testing. It exceeds what has been possible before for testing systems under real-world conditions. This approach merges the real-world traffic generation and measurement capabilities of TDM-side solutions with the real-world data network emulation capabilities of data-side solutions. The TDM tester initiates calls while controlling how the data test platform functions as a network emulator, delaying and modifying packets as they pass through to emulate actual network impairments. No Quality of Service (QoS) can be thought of as the minimum parameters a network must maintain in order for acceptable quality voice to be attained. IP networks were designed with one purpose in mind: to carry "bursty" types of data over flexible paths, with varying amounts of bandwidth, and to connect a wide range of sites all sending and receiving data simultaneously. As such, IP networks have no built-in guaranteed quality of service (QoS). Voice is delay-sensitive and prone to quality deterioration if network parameters exceed limitations. The complexity of the IP network also became a factor as software installed on gateways, routers, and other equipment can manipulate characteristics of the packets that travel through them. What was needed was a new method of testing.

From a single logical test system (physically one or multiple boxes), the user may set test parameters, monitor ongoing results, and review integrated reports documenting corresponding events on both sides of the gateway. This test approach involves a test bed that truly replicates real-world conditions and thresholds and avoids the cost of placing the actual components in the lab. It also avoids time-consuming configuration and reconfiguration of each individual component in order to mimic parameters of implemented networks. This approach monitors Quality of Experience (QoE). QoE is a new term describing the emerging reality that what ultimately matters in moving to the next-generation network is how users think it performs. QoE is not a metric per se, but rather a concept comprising all elements of a subscriber's perception of the network and performance relative to expectations. The concept applies to any kind of network interaction.

From an end-to-end perspective, while also validating that QoE will hold up under real-world network conditions, users determine the profile by which TDM-based calls are generated (media type, call frequency and duration, offset) and, at the same time, define the network packet-handling behavior (jitter, packet loss, duplication, re-order, corruption) for a particular test. Running a test with no impairment, and then running that same test with different network degradation scenarios (all operating at wire speed), while observing the effect on voice quality and latency gives a clear picture of how an integrated solution will behave when deployed. This test methodology is particularly useful for uncovering defects that may not show up in traditional homogeneous approaches. For example, by applying step changes in a particular network condition (impairment), users can determine the point at which nonlinear voice quality scores degrade beyond tolerance.

Besides test configuration and setup, monitoring is another critical element of converged testing. The ability to observe real-time information about a particular test in progress on both sides of the network (for example, calls originated versus calls terminated, and latency or speech quality measurements over-laid on call volumes) is necessary for efficient testing. Monitors allow users to quickly predict thresholds and targets that can be refined over time with more deterministic testing.

Clearly, a Converged Network Emulation approach provides a more realistic assessment of Quality of Service (QoS) than traditional approaches alone are capable of delivering. Although diagnosing and isolating problems can still be a challenge with this approach, converged analysis and reporting can be great aids. Information about specific calls originating on one side of the new world and terminating on the other side can quickly point users in the right direction and can suggest experiments for determining problems.

Data-side test solutions bring some strengths to testing converged networks. They generally support high data rates, which are critical to stressing data networks. They are also capable of detailed protocol analysis to ensure proper data handling. Data test equipment providers have evolved strengths in protocol analysis to apply to a new set of protocols. Recognizing a heightened need for management of QoS in converged networks, they have offered packet-level QoS monitoring tools. Generally, data solutions offer good analysis of activity points in the network. A new level of difficulty arises in converged networks due to the increased number of handoffs from signaling protocol to signaling protocol. To track down handoff problems requires assembling point solutions and piecing together diagnostics from the different sources.

Another converged testing approach is Isolated Device Testing. It simply means "wrapping around" a device or perhaps small set of edge devices and directly testing it. This approach has traditionally only been possible for homogeneous devices. This is the way routers are typically tested in the data world, and it has been used for testing voice devices in TDM networks as well. The challenge for converged networks is testing the heterogeneous gateways on the edge between TDM and data devices. This approach requires (again) a logical test system that can interact with both sides of the gateway and can create realistic conditions for each.

A comprehensive Isolated Device Test solution can uncover issues that would impede QoE of the systems well before they are even close to deployment.

Testing devices in an isolated fashion enables quick testing early in the design process. It provides control and visibility to the isolated device. Thus, a converged solution that is capable of generating TDM traffic and terminating IP-based media streams (and vice versa) from a single interface overcomes many of the limitations of back-to-back testing. Straightforward call load and path confirmation are the obvious benefits. For such a solution to provide a measure of QoE under real-world operating conditions also requires voice quality and media (tone, fax, modem) verification. If the test system supports the ability to compare the PSTN signaling messages to a device with the associated IP messaging then signaling handoff problems identification and resolution is accelerated. For equipment manufacturers this translates into a time-to-market advantage. For solution or service providers testing the interoperability of equipment from multiple vendors, it eliminates the finger pointing that sometimes accompanies those installations.

For all the benefits of the mentioned converged testing approaches, there is a further step in test technology that combines the benefits of both. The Complete End-to-End approach provides the QoE/QoS perspective and real-world nature of Converged Network Emulation and the problem-solving strengths of Isolated Device Testing. This methodology retains the notion of a single logical test interface and expands the visibility and control by placing smart data probes between each device of interest. From this vantage point users will not only be able to control the data backbone and observe the effects on QoE/QoS, they will also be able to trace the hops and conversions as media and signaling traverse these next-generation networks. Device isolation is taken to the next level and pinpointing problems within these vast networks is more easily attained. A test approach of this type will provide engineers access to extremely valuable diagnostic information. For example, after traditional signaling is verified for correctness upon conversion from the PSTN to the packet network through the edge device (probably SS7 to SIP), bearer traffic could be assessed for voice quality. At the same time QoS metrics, such as jitter, packet loss, and bandwidth utilization, could be garnered. QoS-enabled routers and devices would then be measured against SLAs as packets continued to traverse the network. Interdevice messages, such as MGCP, could also be validated between signaling and media gateways. This process would continue across multiple hops until the media stream reached its destination within the packet-switched network or back into the PSTN world. End-to-end assessment would include time and quality measurements, as well as interdevice statistics and overall QoE.

The value of the approach is clear; however, the implementation challenges and the breadth of technical expertise required make this a formidable challenge. The extensive number of handoffs and conversions in next-generation networks (particularly wireless networks) implies long lists of protocols and interfaces. It is unlikely that a single vendor could cover the entire spectrum. Industry standards and open architectures must evolve before the Complete End-to-End Approach testing becomes a reality. The success of next-generation networks ultimately depends on satisfying the high expectations of network users. Testing converged network equipment demands converged testing approaches.

CHAPTER 14

THE NEXT PHASE, APPLICATIONS USING THE CONVERGED NETWORK

Recently, the main driving force behind the deployment of converged networks has been saving money. The theory has been that developing, deploying, operating, and servicing a single network is less expensive than doing all of that in multiple, application-specific networks. This is often quite true. However, with reductions in the cost of items such as network bandwidth and cabling and the increased cost of typical converged networks and devices, this cost savings is sometimes hard to find.

The real driving force behind converged networks is the ability to deploy new applications that take advantage of the converged infrastructure. The unique advantages of a converged infrastructure stem from the fact that devices on the converged network can access voice services as well as traditional data services simultaneously. This allows these devices to use these combined services in new ways.

In this final chapter we will discuss many of the applications that are driving the deployment of converged networks. Some examples of these applications are Converged, Multichannel Contact centers, Unified Messaging, and Teleworkers. There are also new applications under development that will utilize the unique features of the converged network such as presence, intelligent agents, and "IP Centrex".

We will discuss each of these topics in detail.

Contact Centers/Customer Relationship Management (CRM)

One of the most popular converged applications that takes advantage of the capabilities of the converged network is the Multichannel Contact Center. A multichannel contact center expands on the traditional telephony contact center by adding items such as email, web access, and video access streams into the contact center. The multichannel contact center also provides for tighter integration between the customer-facing agents and back-office databases and functions. This can give a business a distinct advantage by giving their customers many different ways in which to contact them.

New technologies that support Internet applications, telecommunications, and knowledge management have resulted in a new, unified approach to managing customer relationships through the integration of enterprise resources into a seamless CRM solution. This unifying solution, which embodies a distributed contact management capability, is known as a multichannel contact server. This contact server provides a platform that enables the integration of business operations, business management, and business intelligence capabilities. It is driven by business rules and processes so as to meet the major objective of providing superior and consistent customer service.

The multichannel contact server

- Interacts with front-office and back-office applications.
- Manages customer contacts from a variety of media and communications channels.
- Routes contacts to appropriate resources based on predefined business rules.
- Integrates information from corporate databases to support the customer interaction.
- Broadcasts the real-time status of available resources to other applications.

Multichannel contact servers are facilities that pull all of these capabilities and resources together around business processes, making them easily accessible for both front-office and back-office uses. Figure 14.1 illustrates the major components of a multichannel contact server, linking communications channels, customers, and back-office operations with customer and business intelligence.

The contact server brings all of the necessary knowledge and capabilities together in the context of specific business processes that are driven by flexible business rules. With a contact server in place, the enterprise can bring its collective resources to bear on each event to ensure a consistent, personalized customer experience. It enables the collection of information about each transaction, and improves customer relationships during future customer interactions.

The contact server is fundamentally a software platform that integrates front-office, eCommerce, and multichannel contact center applications into one centrally managed system. The contact server supplies the foundation for an effective CRM strategy by allowing enterprises to communicate with their customers regardless of the contact type, and to manage the end-to-end business

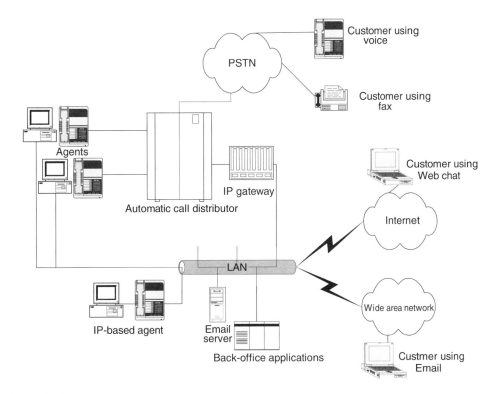

Figure 14.1

A multichannel contact center.

processes necessary to interact efficiently and profitably with customers. It manages customer contacts from a variety of media in a consistent fashion, routes them to appropriate resources based on predefined business rules, integrates information from customer databases into the customer transaction, and broadcasts the real-time status of available resources to other applications.

The integration of contact center resources with front-office and back-office systems extends computer telephony integration (CTI) capabilities such as data-directed routing, "screen pop," and coordinated data transfer to transactions performed via media other than telephony. Customer case histories, account and credit information, inventory data, shipping information, and much more can be instantly and automatically available to both automated and human resources at the exact moment the customer makes contact, no matter what channel the customer uses.

This conceptual architecture supports an enterprise-wide approach to business operations, management, and intelligence. With it, all channels are supported within a single, integrated information framework to ensure high levels of data integrity, quality, and usability. With the contact server, front-office and back-office applications are integrated into the contact server directly, through operational data stores (ODS), through data marts, and through direct data interchange between the underlying components. The contact server becomes the place for business users to manage, execute, and monitor every business process that involves customer interaction.

As with traditional enterprise portals or Web portals, the contact server is more of a place to bring together the capabilities of the enterprise than a place to build capabilities. With the contact server, the context of these groupings of capabilities is the business process. The business rules engine acts as the brain of the contact server, bringing everything else together and managing customer interactions within the other components of the architecture.

Multichannel Support

In most businesses, communication channels can be categorized as follows:

- *Voice:* Telephone, fax, interactive voice response (IVR), and related media.
- *Data:* Computer-to-computer traffic—Web email, chat rooms, forums, and other Internet media.
- *In person:* Sales calls, seminars, trade shows, deliveries, and service calls.
- *Business partners:* Distributors, resellers, consultants, and others.
- *Direct mail:* Catalogs, order confirmations, bills, deliveries, and marketing messages.
- *Mass media:* Television, radio, magazines, newspapers, and other mass media vehicles.

The contact server must consistently align these varying channels around business processes. Contacts in some channels occur in real-time, with customer contacts taking place in highly interactive media such as telephone or Web. Others may occur over a day or more, such as email and fax. Clearly, real-time customer interactions are the most challenging to manage and integrate.

The capabilities supported by the contact server can be categorized as dealing with contact center channels, back-office and front-office operations, as well as collaboration, business management, and business intelligence. The contact server wraps around all of these capabilities and enhances them with the business rules engine to drive specific business processes and knowledge-based applications.

Standards in the Contact Server Architecture

In contemporary networking, it is important that systems be built on open standards and architectures in order for enterprises to leverage their existing investment in their infrastructure and applications. The contact server architecture is based on industry standards as well as solid de facto standards to ensure long-term viability. It enables the interoperability of the contact server with current and future technical components used to support CRM.

Due to the key role of multichannel contacts in most enterprises, technical interoperability extends beyond just the information systems to the telecommunications systems. The need for interoperability is especially critical in the CRM arena because of the rapidly evolving and consolidating field of diverse technologies used by companies to interact with their customers.

One need look no further than the new generation of personal wireless devices, voice over IP, and Internet-ready home appliances and vehicles to get a glimpse of the importance of applying industry standards to a contact server. This is essential if the contact server is to continue to be successful over the long haul.

Some key standards include

- *ODBC (Open Database Connectivity)*—ODBC is a widely accepted API for database access. It is based on the call-level interface (CLI) specifications from X/Open and ISO/IEC for database APIs and uses structured query language (SQL) as its database access language.
- *XML (Extensible Markup Language)*—A key tool in the enterprise contact server is the extensible markup language, XML. XML is a clearly defined set of rules for a common communication language on the Web. Through XML, various industries have been defining standard documents to use within their community (purchase orders, invoices, etc.), and it has become a favorite for integrating heterogeneous and distributed systems. XML enables the minimum representation necessary for message and data exchange between different components or bodies. The core specification is extremely simple and has proved itself to be extensible, robust, and flexible. While the Internet is always going to be made up of different operating systems running on different platforms using applications written in different programming languages, XML provides the basis for them to seamlessly interoperate in a distributed environment. XML was created so that richly structured documents could be served, received, and processed over the Internet. It maintains the separation of the user interface from the actual data, allowing for increased flexibility and ease of administration. XML is used to describe the actual content and structure of a document and individual applications interpret how the document is to be viewed. HTML (hypertext markup language) and WML (wireless markup language) are used to display the information held in XML documents.
- *ActiveX*—ActiveX is a suite of software technologies that enable components to interact with one another in a networked environment, regardless of the language in which the components were created. This is done through

a series of ActiveX user interfaces called ActiveX controls that are small, fast, and powerful. These controls make it easy to integrate and reuse software components in a Windows environment.

- *CORBA*—The dominant distributed object technology used on the Web today is the Common Object Request Broker Architecture, CORBA. CORBA is an open industry standard that provides an object-based middleware to support different platforms. It has proven to be an effective mechanism for system integration and for providing Internet access. CORBA is used in UNIX-type environments and interfaces with JavaBeans.

- *JavaBeans*—Sun published the first version of its Enterprise JavaBeans (EJB) specification, which is a Java-based, component-oriented framework for developing, deploying, and managing distributed, transactional applications. CORBA, with its services and EJB, addresses the issue of application development across multiple platform types. Although the two specifications have been developed independently, it turns out that the two technologies are complementary. These software solutions work in conjunction with HTML, dynamically generated either by servlets or Java Server Pages (JSP). The middleware center of the contact server controls business logic and provides access to various back-end systems, including relational and object databases, application software packages, and legacy systems.

- *VoIP Gateway H.323 and SIP*—A key capability of the contact server is to integrate voice traffic into the full set of Web-based interactive functions. This is accomplished through a set of standards that operate on various gateways, application servers, and switches. Among the most important are H.323 and Session Initiation Protocol (SIP). These provide the translation services from voice to data, so that voice traffic can be integrated into the IP-based traffic stream.

Functional Concerns of Multichannel Contact Servers

A successful contact server includes several important functions, as follows:

- *Scalability*—As the nerve center to customer interactions across the enterprise, the contact server must be based on a sound and flexible technology architecture to ensure its ongoing success. The architecture must be scalable, meaning that it must be able to grow with the enterprise without discarding the existing technology investments. Aspects of scalability include growth of processing power, growth of memory, growth of disk storage, and growth of network from both data and telecommunications perspectives.

- *Enterprise-Wide Perspective*—CRM requires an enterprise-wide perspective, and optimally supports customer contacts across all possible communication channels. Likewise, the contact server must support all customer-facing functions, including marketing, sales, fulfillment, and services, as well as some cross-functional capabilities such as risk management. Interactions involving customers, partners, or the supply chain should be managed to ensure a consistent experience for the customer and a profitable long-term relationship for the enterprise.

- *Interaction and Response Templates*—The business process should support interaction and response templates. An example of an interaction template is a contact center script that directs an agent through an interaction step by step. A response template may be a template email that is easily customized by an automated system, a salesperson, or a contact center agent to respond to a certain type of request.
- *Rules Engine*—In order to support a flexible, business-process-oriented capability, the contact server must have a robust rules engine that executes processes and makes independent decisions based on rules defined by business analysts. These business rules provide routing of interactions across channels and functions, provide appropriate scripts and response templates, and personalize the customer experience based on individual customer profiles.
- *Integration Functions*—These combine platform middleware functions with the ability to transform data and route it according to a set of rules, enabling peer-to-peer real-time integration between applications, including enterprise resource planning. In a sense, it is a "push" technology moving data quickly to a CRM application so it is always available for use. The integration solution relies on an approach to data management where data is classified and treated according to its classification.
- *Contact Classification and Prioritization*—There are many different types of customer contacts. A customer interaction may be a product order, a request for information, a problem report, a product registration, or an invoice payment. Though these contacts are given different priorities, before they can be prioritized, they must be classified. The contact classification, in conjunction with the customer profile and the available staff, allows the business rules engine of the contact server to properly route and prioritize each contact with the enterprise. The priority of each contact is then based on the importance of the contact to the customer, the importance of the contact to the enterprise, and the relative value of the customer relationship. Even the time of day may result in different priorities for the same contact. The role of the contact server rules engine is to properly prioritize based on the customer profile, the classification of the contact, and any other information to which the business can apply a rule.
- *Customer Data*—Customer profile information contains attributes of the customer such as name, address, date of birth, and so on. It is defined in the customized database schema of the CRM application. In the CRM solution, this data is under the control of the CRM application, and it can only be changed by the CRM application. Any changes made to the customer data by the CRM application will be made available throughout the system as needed. Specific customer attributes hold information about the customer, based on the real-time analysis performed in the business processes. These include attributes about customer value, their propensity to continue to

engage, relevant products to attempt to cross sell, and so on. These values are generated in real time and as a back-office function.

- *Minimum Product Data Sets*—This is the minimal set of product data that the CRM application requires for interactions with the customer. It is often stored in a customized area of the CRM application and can be changed by the CRM application. The contact server will propagate any changes made by the CRM application to the minimum product data set.

- *Extended Product Data Sets*—This is additional information about products or services from the underlying core business systems that is accessed by CRM application. This is a compromise between the size of the CRM application database and the ability to function efficiently. The intent is to ensure that the bulk of customer transactions are handled by the minimum product data set. This extended data is usually not stored permanently within the CRM application but in a customized area of the CRM application used for staging purposes. Attempts to access this data will cause it to be fetched in real time from the relevant system and stored temporarily in the CRM application. Any changes made will be propagated back to the core system.

The multichannel contact server solution represents a proven technology that meets the challenges of our rapidly changing business environment. The contact server plays a key role in integrating eCommerce and the Web with traditional customer interactions. The emergence of the Internet has done more than change our technology and our business. It has fundamentally changed the entire global marketplace and society as a whole.

Customer expectations are continually rising. With products and services now only a click away from hundreds of sources, retaining customers requires more than simply doing business as usual. It requires attention to individual customer relationships in a consistent and efficient manner across the enterprise. Depending on the industry and markets, the contact server may provide competitive advantage, increase revenues, and streamline interactions with individual customers—or it may simply mean survival for the enterprise.

The multichannel contact server integrates the front-office, eCommerce, and multichannel contact centers into one centrally managed eBusiness system. It allows media blending, workforce management, and customer self-service. It brings the power of business intelligence to bear for increased efficiency and customer satisfaction. It reduces application development time, cutting development costs and realizing quicker return on investment (ROI), allowing the enterprise to respond rapidly to changing business conditions.

The contact server becomes the view of the enterprise into this capability to maximize the value of each customer relationship. The rules engine serves as the nerve center, directing traffic based on rules-based workflow, moving data from one point to the next as the customer interacts with the enterprise. The emergence of Web is the primary catalyst behind the need for capability because now there are two highly interactive, diverse means of communicating with customers—the telechannel and email. Coordinating these two sets of communications media in the context of ever-increasing customer expectations for consistency and dependability requires more than yesterday's tools.

The consistency expected by today's customers requires a means to bring together a diverse set of business strategies, people, and technologies in a constantly changing marketplace. This is the role of the multichannel contact server.

Unified Communications/Messaging

Another key application that is evolving to use the power of the converged network is the concept of Unified Messaging. Unified messaging is a generic term for a set of applications that links all voice mail, e-mail, fax, and other messaging types into a single coherent messaging platform.

Effective communication is critical to the success of businesses today. Messaging has become a key component of corporate communications, allowing organizations around the world to communicate across distances and time zones easily and efficiently. Workers today constantly check their voicemail via telephone, their email via PCs and laptops, and their faxes on various fax machines. However, these disparate messaging systems may evolve into isolated information "islands" that add to workers' stresses and actually erode their efficiencies. For mobile workers and telecommuters the situation is even more serious because not all message types can be accessed by telephone, the only ubiquitous terminal device available. Most companies with multiple locations around the world rely on a number of different messaging systems to support multiple languages as well. In addition to end-user problems, it is also difficult for information technology departments to manage and provide customer support for several different messaging systems.

Unified messaging is a new technology that unites disparate systems into a single, unified mailbox and enables access from the PC or from any touch-tone telephone. This streamlined access to and management of information dramatically enhances the productivity and responsiveness of office workers, telecommuters, mobile employees, and IT staff throughout an organization. The people within the enterprise are better equipped to make fast, effective business decisions in the office or on the road.

A unified system brings together voice, e-mail, and fax-messaging systems—and in the future, video-messaging applications—to allow a user to access all messages from a single unified mailbox via a PC or telephone. This system also supports user-customizable rules-based access that allows users to flag and prioritize important messages by subject or sender. This capability can be particularly useful to mobile professionals who need quick access to their messages, anytime, anywhere, and from the most convenient device.

Technical Overview

Companies have a choice of architecture that best suits their enterprise topology. A unified architecture allows system administrators to manage a single scalable and converged voice/data messaging system with a single message store.

Alternatively, a company may wish to deploy unified messaging via an integrated architecture with multiple message stores.

In small organizations, or for limited implementations in larger organizations, these systems can be configured to provide comprehensive voice, fax, and email messaging services all on the same server (see Figure 14.2).

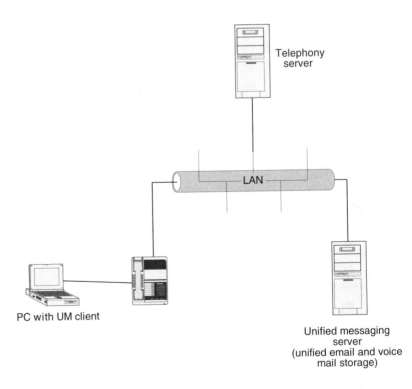

Figure 14.2

Unified messaging on a single storage device.

For larger installations, these functions can be divided up among separate servers. This enables customers to take advantage of in-place resources, such as a common message store or global address lists. It also allows enterprises to distribute their unified message store among their existing email servers, or use the unified messaging server to store a subset of user messages (see Figure 14.3).

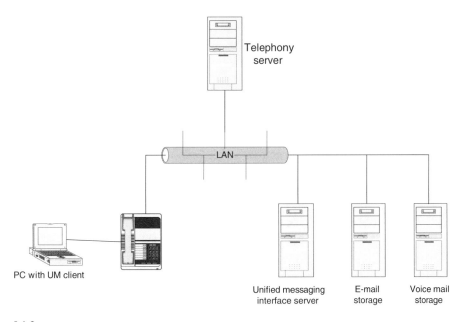

Figure 14.3

Unified messaging with separate voice and email storage.

High-end systems will also provide a single directory service that simplifies both message addressing as well as system administration. Many of these high-end solutions integrate the messaging components to each other and to the rest of the network services using the Lightweight Directory Access Protocol (LDAP) services.

End-User Access

End users can use mail clients, such as Microsoft Outlook or Lotus Notes, to visually access voice, fax, and email messages. Unified messaging systems augment users' email clients with custom forms that enable voice messages to be played and recorded, and faxes to be created, viewed, and printed.

In most unified messaging systems, users can also access their messaging using a telephone user interface (TUI). The TUI gives access to messages via DTMF (touch-tone) commands. Many systems can support test-to-speech services to deliver text-based email messages over the TUI. Some newer systems even allow for access to mail via voice commands.

Voice messages are easily played and recorded from a multimedia-enabled PC, but users that don't have multimedia PCs can still leverage the systems multimedia capabilities. Many systems also allow the user to re-direct the voice playback to a phone number the user specifies. Then a user can use the telephone to play and record voice messages, while still managing messages with the convenience and flexibility of an email client interface.

The system may have a web-based personal administration interface, which enhances productivity and message control by providing an easy-to-use web interface managing mailbox functions. Advanced features such as outcalling schedules and personal distribution lists can be managed easily using the web interface, making these features much more readily usable for all users.

System Administration

System administration for the unified messaging application often uses a centralized applications server with either a client- or web-based interface that streamlines installation, maintenance, updates, and other system administration tasks, minimizing the need for separately administering the discrete system components. Remote system administration can be supported as well. A unified messaging system often also takes advantage of LDAP services, which provide a standard enterprise-wide format for storing, exchanging, and updating user information. Instead of constantly recreating directories, system administrators can use LDAP services to update directories throughout the enterprise. For example, customers that integrate their system with an Exchange network can use the Exchange Directory service to perform directory maintenance and replication across sites. Lotus customers can have this same access in their Domino Domains.

Unified Messaging Applications

With unified messaging, workers have flexibility in managing their voice, fax, and email messaging. Common applications for Unified Messaging Include

* *Converged Message Management*—Workers can access all their incoming messages from one inbox using either a PC or a telephone. Workers can

exchange voice messages from their desktops using PCs. Remote and mobile workers can access their fax and email messages via the telephone using integrated fax store and forward capabilities, which allows subscribers to direct stored faxes to any convenient fax machine.

- *Messaging Morphing*—Many workers have some difficulty dealing with messages sent to them in a non-preferred format. Unified messaging greatly reduces this problem by letting users choose the most appropriate medium to receive and reply to their messages. For example, a user might receive a voice message but choose to answer it via email. Or a user might receive an email message and decide to forward it to another user as a fax document. Users can also use the telephone to have their email messages read to them by using a text-to-speech application. Some systems even allow users to blend message formats and create "compound messages" that combine voice with fax or email.

- *Intelligent Agent*—A unified messaging system can feature an intelligent call-handling feature that enables personal auto attendant applications. These applications allow users to configure a custom menu of routing options for callers when they reach a subscriber's mailbox. A typical application may give callers options to reach an assistant, try a cellular phone, send a fax, or leave a message. Personal auto attendant makes it easy for callers to reach subscribers through a variety of methods from a single telephone number, eliminating the need for users to give out multiple phone/pager numbers.

- *Personal Assistant*—Using a web browser screen or client application, individual users can configure different notification options to a variety of devices to increase their ability to be alerted of critical incoming messages. For example, by using a daily schedule set by each user, users can be notified of specific message types with a certain level of importance, to whichever device they choose. This feature can be made accessible via any touch-tone phone as well as via a browser screen or client application.

- *Fax Management*—Fax management capabilities allow users to store, forward, and reply to fax messages just as they do voice and email messages. Instead of being printed at a fax machine, fax messages are stored in the subscriber's unified mailbox. Subscribers decide where and when to print fax messages, ensuring privacy and security. Mobile workers can use the telephone to send fax messages or email messages to a local fax machine when it is convenient. This functionality gives workers easy access to incoming faxes even while traveling, again keeping them in touch with customer or co-worker requests for information. The ability to also print any email to the nearest fax machine further enables mobile workers to access key information received as email, even while not sitting at their computer. Fax messages can also be annotated with voice comments using compound messaging capabilities.

Teleworkers

Probably the one application that most enterprises have embraced for their first converged application is teleworkers. The teleworking applications allow users to access voice functions over their already-existing remote IP access applications.

Large enterprises in particular are faced with an increasingly dispersed workforce. Current market research shows that only one-third of enterprise employees actually work in main office locations. The majority of this market's workforce is employed in branch office and remote locations. As much as 16 percent of the entire enterprise workforce currently telecommutes at least three days a week. As companies seek to remain competitive in local and global markets, further fragmentation is expected to increase the number of remote workers, especially the percent of teleworkers, making up the enterprise workforce.

In today's economy, intellectual capital is a key business asset, and the value of knowledge workers continues to grow. Regardless of the state of the economy, employers are challenged with hiring and keeping quality employees, particularly in the high-technology sector. As a result, the best and brightest of these potential employees can dictate their own terms, and may not want to relocate or even commute on a regular basis. If companies want their services, they may have to employ them where they are.

Clearly, there is increasing awareness of and interest in teleworking throughout the labor pool. And these would-be teleworkers have the government on their side. The federal government recently mandated that teleworking be made available to a portion of federal employees whose jobs are suited to it. Many local governments are using tax breaks and other incentives to motivate businesses to embrace teleworking as a way to reduce congestion and air pollution in population centers.

However, market forces are much more compelling. Forward-looking companies recognize that teleworking is a win–win proposition for employer and employee. With a good teleworking solution in place, corporate facilities don't have to be over-provisioned to meet peak demands. Companies can recruit the best people from anywhere, and don't have to limit themselves to the local talent pool.

The most notable benefit of teleworking can be measured through personal productivity gains. There are countless studies that suggest employees who work from home outperform in-office workers by as much as 15 to 40 percent. This is due to uninterrupted work time, better concentration, and longer work days. Eliminating a commute on telework days not only reduces stress and fatigue, but can also add one or two hours of productive work time to the day.

In addition to productivity gains, offering telework as an option helps to attract and retain employees. Businesses with a telework option experience less employee absenteeism and greater employee retention. Both employers and employees can also realize significant savings on real estate spending. Workers can get more home for their money by locating farther from the office in areas with lower real estate prices, and businesses can employ an "office hoteling" model that saves space on corporate campuses by eliminating the traditional 1:1 ratio of employees to desks.

Teleworking can also enable large companies to extend their presence into regional and local markets without the costs of opening a branch or satellite office. By locating customer-facing employees such as sales representatives, customer support personnel, and insurance claims adjusters in the local community, companies can provide better customer service, increase revenue, and stay one step ahead of their competitors.

Besides these tangible savings, teleworking benefits also include reduction in air pollution and traffic congestion. This establishes the company as not only supportive of its employees, but of community and environmental concerns as well.

Early teleworking solutions relied on remote data access, and did not integrate the remote worker into the corporate communications infrastructure. Teleworkers were viewed and treated as part of a different community of workers. This bred isolation and caused management challenges. Legacy PBX suppliers began extending corporate voice applications to the teleworker, but they often had poor voice quality and reliability, and were too expensive to implement and manage on a large scale. On the data side, teleworking solutions have made great strides, especially since the advent of broadband network access. Many people can now get high-speed DSL or cable access that can support media-rich content and real-time collaboration.

The IP-based converged network can exploit the growing ubiquity of inexpensive broadband access and extend the corporate voice and LAN infrastructure to remote employees. By using an IP remote access scheme such as dialup or VPN, the user can gain access from a remote site to the local area network. Once the remote user has access to the LAN, that user can then access both the data applications and electronic messaging applications, and also access the voice media servers and messaging as well. The teleworker can deploy an IP voice endpoint or a converged voice/data endpoint (for example, an IP softphone) to access the functions of the converged network (see Figure 14.4).

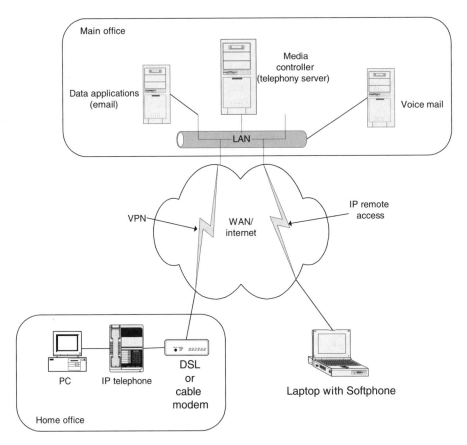

Figure 14.4

Example teleworker configurations.

There are several user types that can benefit from the deployment of teleworking applications.

Executives. Managers who work at home only occasionally, but need full, secure, real-time voice and data connectivity can get this access of the remote access link.

Key Knowledge Workers. Highly skilled people are often difficult to find, and the company sometimes needs to be brought to them if it is to avail itself of their services.

Occasional Teleworkers. Rank-and-file office workers who want to work at home occasionally. can access full office functionality over the remote link.

Mobile Employees. Employees who travel a great deal or whose job functions make them work out of the office can also access full corporate voice and data services from a remote access link.

Call Centers: Teleworking is having a major impact on call centers. As technology products become increasingly ubiquitous and commoditized, customers take for granted their complexity and expect instantaneous response from call center agents. As a result, call center employees are being forced to evolve from phone operators to full-fledged knowledge workers.

Experienced call center employees are hard to find and even harder to hire. Allowing them to operate from home, as part of a virtual call center, may be an effective solution to the labor shortage problem. This means extending call center stations to the employee's home with a teleworking solution that delivers the same functionality, reliability, and security of a seat in a traditional bricks-and-mortar call center.

By moving key call center components to open systems on the IP data infrastructure, you can build a very cost-efficient solution that is easy to operate, distribute, and scale to accommodate peak demands and future growth.

The remote location of the agent can be made completely transparent to the callers. By extending the control of the call center over a remote link such as VPN, the call center measurement and reporting tools can also monitor and support the remote agents in exactly the same way as the on-site agents.

Executives, call center professionals, and key knowledge workers need the full corporate campus voice functionality and experience extended into remote sites. Deploying IP-based solutions has made such high-end teleworking affordable and easy to implement.

SIP, Presence, and Intelligent Agents

Yet another one of the new applications being developed for the converged network stems directly from the development of the Session Initiation Protocol (SIP). SIP is a signaling protocol for establishing calls and conferences over IP networks. The session setup, change, or tear down is independent of the type of media that will be used in the call; a session may include different types of data, including audio, video, and many other formats. Those who are most excited about SIP believe this development is as significant as the HTTP protocol, the technology behind web pages, that allows a single page with clickable links to connect you to text, audio, video, and other web pages. This possibility is behind SIP's rapid adoption as a Voice over IP standard. The protocols supporting SIP are discussed in more detail in Chapter 9.

SIP is modeled after other Internet text-based protocols such as SMTP and HTTP, and was designed to establish, change, and tear down calls between one or more users in an IP network in a manner totally independent of the media content of the call. Like HTTP, SIP moves control over the application to the endpoint, eliminating the need for a central switching function.

SIP Architecture

Main components of the SIP architecture are

1. SIP User Agent—The User Agent is the SIP endpoint or end-station software. The User agent functions as a client when initiating session requests, and also acts as a server when responding to a session request. Thus, the basic architecture is client/server in nature. The User Agent is "intelligent," in that it stores and manages call state. The User Agent places calls using an email-like address, or a telephone number (E.164). As an example: SIP:user@university.edu. This makes SIP URLs easy to associate with a user's email address.

2. SIP Server
   a. SIP Proxy Server One type of SIP intermediate server is the SIP Proxy Server. Proxy Servers forward requests from the User Agent to the next SIP server, and also retain information for accounting/billing purposes. In addition, the SIP proxy server can operate with stateful (i.e., circuit-like) or stateless (i.e., TCP-like) communication. The stateful SIP server can "fork" incoming calls so that several extensions are rung at once and the first to answer takes the call. SIP proxy servers can use multiple methods to try to resolve the requested host address, including DNS lookups, database lookups, or relaying the request to a "next" proxy server.
   b. SIP Redirect Server—A second type of SIP intermediate server is the SIP Redirect Server. The Redirect server responds to the User Agent's request by providing information about the requested server's address so that the client can contact that address directly.

   The role of these SIP servers is to provide name resolution and user location. The combination of Proxy and Redirect servers gives SIP great architectural flexibility; the user can employ several schemes, simultaneously, to locate users. The SIP architecture is particularly well suited to support mobility.

3. SIP Registrar—The SIP Registrar provides a location information service; it receives information from the User Agent and stores that registration information.

The SIP Architecture allows for the support of a number of new, intriguing applications on the converged network. These new applications are built based on a concept called "presence." Presence is the combination of the User Agent along with the functionality of the SIP proxy and redirect servers as well as the SIP registrar to build information on the status and preferences of a given user.

SIP Presence services can be used in many applications.

Specialized "call forwarding" allowing users to specify where they are so that incoming calls can be forwarded there, or to choose to forward calls to "voice mail" or any other answering service.

Presence, in conjunction with translation gateways, can even be used to alter the message type. For example, a message could be taken in email, but based on the user's preferences the message could be translated into a web instant message and sent to the user's terminal.

Call participants can manage the call; this allows one or more callers to decide to bring in a new call participant or to cancel a connection in the call.

The user has the ability to return different media types. For example, allowing an incoming call to be answered by a web page with information that can be used to complete the call.

The User Agent can be used to indicate whether the user is present (available to take the call) or absent (not able to take the call). The user agent could also be used to determine where a call should be routed, so a call that is intended for the office could be re-routed to a cell phone based on preferences stored in the user agent.

The Presence information can be used for a wide variety of exciting new services, such as

- Establish an Instant Messaging text chat session,
- Make a phone call when the called party becomes available,
- Invoke an instant conference when all the desired parties are online,
- Get a notification when a cell phone is on the air,
- Get a notification if an agent becomes available,
- Specific vertical applications, such as monitoring delivery vehicles or employees on the move.

Notice that SIP-based Presence and Instant Messaging can be supported not only on PC/laptop computers, but due to the small GUI footprint and low computational requirement, can also be supported on intelligent SIP phones, palm computers, and even mobile phones.

Many desktop software vendors have great expectations of SIP and presence being tied into their applications. A good example of this is to tie the user agent into a calendaring software package, so that calls can be re-directed automatically based on entries in the calendar.

SIP and the concept of "presence" may well be a tool that redefines a lot of what we are used to in modern communications.

IP-Based Public Services

IP Network Access

Converged network services are not exclusive to the enterprise network. Network service providers are also looking very closely into converged networks for supporting their customers as well.

All of the features and applications that we have discussed throughout this book could equally apply to service provider-based networks. In fact, there are several carriers that are providing converged services such as Voice over IP network access today.

By using VoIP technology, service providers can give their customer direct voice, data, and video access to their network backbones. One of the biggest advantages of this is to give very high bandwidth (10 Mbps to 1 Gbps) access at a relatively low cost.

By providing a converged access to their network, service providers can not only gain the advantages and cost savings of combining voice and data access connections. They can also build converged applications in their network and sell these as value-added services.

Network service providers are also some of the strongest supporters of the SIP architecture. This is because SIP gives a simple standard interface to converged services. Because of the simplicity of SIP, carriers see this as a way to support simple, low-cost, vendor-independent VoIP terminal devices as well as advanced network services such as

- Global ENUM service for phone number to Internet address conversion,
- Global presence on IP, PSTN, and mobile networks,
- Global roaming,
- Global call routing,
- Gateways to the global PSTN,
- Inter-enterprise conferencing.

All of these services can be applied across the entire public network infrastructure and give users access to advanced business features regardless of location. The service providers also enjoy new opportunities in new markets.

IP "Centrex"

A major area where public carriers are using VoIP technology is for providing enterprise telephony services from the carrier network. This is essentially a redefining of a service that has been around for many years, known as "Centrex."

Centrex or Centranet, as it is sometimes called, is a service where a carrier provides functionality similar to a business PBX over a public network interface. The idea behind Centrex was to provide advanced features such as extension dialing, call forwarding, hold, call transfer, and messaging services to customers without those customers having to deploy on-site telephony equipment.

The issue that often stifled Centrex deployment were that user features and the protocols to support them were not standardized. That led to Centrex having fewer user features than a typical premise PBX and also made the service fairly expensive, especially in large implementations. Centrex also had the issue that the users did not control the infrastructure. This meant that changes were typically more difficult and often involved service charges from the carrier.

By using VoIP, and especially SIP, in the carrier's network, carriers can offer full feature functionality over standardized interfaces. By use of presence and proxy servers either at the user's or carrier's location, SIP calling services and features can be managed on a much more flexible basis. Using SIP technology, service providers can offer new, value-added services to their business customers. SIP services such as re-direction, translation, and message forwarding can all be offered to customers as extra options. SIP also holds the promise of relatively inexpensive, standardized endpoints.

IP Centrex could, in theory, deliver all of the benefits and applications of a converged voice/data network that we have discussed throughout this book, without having to support voice equipment at the enterprise's location. This may prove to be a very tempting solution, especially for smaller locations.

Many network carriers fully believe that VoIP and SIP will bring about renewed interest in service provider network-supported business functionality. It will be interesting to see how well IP-Centrex is accepted as it is further deployed in the real world.

Summary

Throughout the course of this book we have mostly discussed the technologies and protocols necessary to design and deploy a converged, IP-based network. While the technology of the converged network is the main topic of this book, it is meaningless unless that new network gives access to new solutions and applications. In the real world, no one will deploy new technology just for the sake of technology.

In this chapter we have discussed some of the major applications driving convergence today. These are applications such as Multichannel Contact Centers, Unified Messaging, and Teleworkers. We also discussed new, emerging applications such as Presence, Intelligent Agents, and IP Centrex.

In the end it is these new applications, along with several others that may not yet be defined, that will be the real drivers of the Converged Voice Video and Data Networks.

INDEX